Complementarity in Mathematics

Mathematics and Its Applications

Volume 1

Willem Kuyk
Professor of Pure Mathematics,
University of Antwerp (R.U.C.A. and U.I.A.)

Complementarity in Mathematics

A First Introduction to the Foundations of Mathematics and Its History

D. REIDEL PUBLISHING COMPANY
Dordrecht-Holland / Boston-U.S.A.

Library of Congress Cataloging in Publication Data

Kuyk, Willem.
Complementarity in mathematics.

(Mathematics and its applications ; v. 1)
Bibliography: p.
Includes indexes.
1. Mathematics–Philosophy. 2. Mathematics–History. I. Title. II. Series: Mathematics and its applications (Dordrecht) ; v. 1.
QA8.4.K87 510'.1 77-8838
ISBN 90-277-0814-2

Published by D. Reidel Publishing Company
P.O. Box 17, Dordrecht, Holland

Sold and distributed in the U.S.A., Canada, and Mexico
by D. Reidel Publishing Company, Inc.
Lincoln Building, 160 Old Derby Street, Hingham, Mass. 02043, U.S.A.

Printed in The Netherlands

To:

Minke, never searing

Denise, most endearing

Editor's Preface

The unreasonable effectiveness of mathematics in science...

Eugene Wigner

Well, if you knows of a better 'ole, go to it.

Bruce Bairnsfather

What is now proved was once only imagined.

William Blake

As long as algebra and geometry proceeded along separate paths, their advance was slow and their applications limited.

But when these sciences joined company, they drew from each other fresh vitality and thenceforward marched on at a rapid pace towards perfection.

Joseph Louis Lagrange

Growing specialization and diversification have brought a host of monographs and textbooks on increasingly specialized topics. However, the 'tree' of knowledge of mathematics and related fields does not grow only by putting forth new branches. It also happens, quite often in fact, that branches which were thought to be completely disparate are suddenly seen to be related. This series of books, *Mathematics and Its Applications*, is devoted to such (new) interrelations as *exempla gratia*:

- a central concept which plays an important role in several different mathematical and/or scientific specialized areas;

– new applications of the results and ideas from one area of scientific endeavor into another;
– influences which the results, problems and concepts of one field of inquiry have and have had on the development of another.

With books on topics such as these, of moderate length and price, which are stimulating rather than definitive, intriguing rather than encyclopaedic, we hope to contribute something towards better communication among the practitioners in diversified fields.

Michiel Hazewinkel

Table of Contents

Preface

This book evolved from a series of lectures presented by the author at McGill University, Montreal, the Free University at Amsterdam, and Antwerp University. These lectures were intended to cater to students of mathematics and students of philosophy simultaneously. As I did not find any suitable introductory textbook in the literature uniting the philosophical, historical, as well as purely mathematical points of view, I decided to venture a set of notes myself. It was from these notes that this book was developed. It is hoped that it contains information about mathematics in non-mathematical terms as well as information on the philosophy of mathematics having meaning for mathematics as a specialist science.

One book which sets out to do the same sort of thing is the encyclopedic *The Foundations of Mathematics* (Torch Books, Harper & Row, 1965) by my former teacher, the late E. W. Beth. I remember some of his classes very lucidly, and with gratitude. He often maintained that if only jurists and politicians, and, more generally, students of any one of the *artes humaniores*, could force themselves to think more in terms of mathematical (and formal) logic, then the world would be a better place to live in. Perhaps so. The point of view of the present book reserves a more modest role for formal method and general mathematical thinking patterns. On this view, mathematical logic and formal methods are specific ways by which the human mind operates in order to unlock the spatial and quantitative aspects of the world. Thus, after having formed, via a complicated learning process, the concepts of continuum (geometric 'entities') and discreteness (the 'entities' of natural numbers), the human mind enjoys a great degree of freedom to operate with them as 'basic material' for the construction of 'structures' such as, for example, particular geometries or groups, rings and fields. Hence, unlocking the spatial and quantitative aspects of the world is as much a learning activity as a creative one. As there is freedom of choice as far as the axioms of structures are concerned, the nature of the basic material as well as information from general experience guide the human mind towards structures that are useful either within mathematics itself or for applications to other domains of investigations such as physics. The formal method can be viewed as a means of making explicit in symbols on paper the logical and mathematical suppositions (axioms etc.) and thinking processes that go into a spontaneous mathematical theory. Thus, it is not impossible that, for instance, certain parts of the juridical science could be formalised. However, in the light of the humanizing tendency nowadays to keep an eye open for the danger of letting computers and other scientific devices reign over man's life, it seems hardly necessary to call formalised applications of juridical theories ludicrous.

There are three chapters in this book. The second one, the longest, dealing with the history of the foundations of mathematics, is sandwiched between the first chapter, describing formalised theories, and the final one, giving an outline of the 'complementarist' view of the author. In Chapter III it is argued that a 'complementarity principle' underlies mathematics as well as physics. This principle, which, in physics, is known by the name 'Bohr's principle of complementarity', enunciates, among other things, that the discrete and the continuous are two different, competitive, aspects of the world viewed from the vantage point of the mathematical and physical sciences (and of the other sciences as well). Accordingly, the mathematical subdisciplines fall into two categories, namely, the category of those which have their epistemological origin and final purpose in the domain of the discrete and of number theory, and the category of those disciplines that stem from (and are intended for) the domain of the continuous and geometry. In the same chapter it is argued, more generally, that there are several aspects (psychological, applicational, set-theoretical, naïve-logical, formal-logical, constructional, etc.) in spontaneous mathematical thinking which are *complementary* to each other in the make-up of that thinking process as a whole; no philosophy of mathematics should ignore these aspects, except at the expense of running the risk of reductionism of some sort or other. Philosophical logicism, formalism and neo-intuitionism are being described as reductionisms of this kind. An axiom system for set theory is given that, hopefully, corresponds as closely as possible to the complementarist view.

In Chapter I an outline is given of Gödel's completeness and incompleteness theorems, which constitute the most incisive results in the foundations of mathematics of the last four decades. There are popularized accounts of these theorems available (cf. E. Nagel and J. R. Newman, *Gödel's Proof*, New York University Press, 1958). We have given an account of all the necessary logical apparatus needed for the Gödel proofs except the Gödel numbering; consequently, we were able to give a description of the final proof itself. It is believed by the author that the general philosophical consequences of these theorems have not yet been fully worked out. Some conclusions from them are to be found in the text. We have also included in this chapter a brief account of non-standard mathematics, which is a very popular subject in the foundations of mathematics today.

In Chapter II we give an account of the history of the foundations of mathematics from Thales to the present, by discussing the work, or part of it, of the leading figures in the field. In doing so we have kept track of how these people account for the different complementary aspects of the mathematical activity hinted at before. We were not able, however, to go into such detail as trying to settle well-known disputes on the interpretation of these figures. For instance, there are different interpretations of Kant's view of mathematics in connection with his analytic and synthetic judgments *a priori* (cf. E. W. Beth, *loc. cit.*). This matter is only touched upon in this text. Hence, the material included should be viewed as first reading material.

University of Antwerp W. KUYK
Formerly at McGill University

Acknowledgements

I am indebted to two students – A. Tol and W. R. de Jong – for taking the first notes of my lectures. The former student was also very helpful in adducing material for the part of Chapter II on Greek mathematics, as this subject was only touched on in the course. I thank Dr M. A. Maurice for drawing my attention to several points that deserved to be included in the final text. I am grateful to Dr Mario Bunge of McGill University for his original suggestion that I convert my notes into a book. Finally, I thank my secretary Mrs M. De Maesschalck-Vleugels for the typing of the often almost illegible.

W. KUYK

Chapter I

Semantical and Syntactical Aspects of Elementary Mathematical Theories

-τοῦ μὲν οὖν μονίμου και βιβαίου καὶ μετὰ νοῦ καταφανοῦς μονῖμους καὶ ἀμεταπτώτους καθ' ὅσον τε ἀνελεγτοις προσήκει λόγοις εἶναι καὶ ακινήτοις, τοῦτου δεῖ μηδὲν ἐλλείπειν τοὺς δὲ τοῦ πρὸς μὲν ἐκεῖνο απεικασθέντος... - Plato, *Timaios 29b*.

[The statements regarding the abiding and the indubitable and regarding that which admits of theoretical reflection, should be constant and unchangeable, even as much as possible irrefutable and solid, and, in this respect, should not let anything be lacking...]

I.1. INTRODUCTION TO THE ELEMENTARY PREDICATE CALCULUS WITHOUT EQUALITY

Our colloquial (natural) language is, scientifically seen, imprecise and our thoughts expressed in it are usually multivocal. For this reason, our colloquial language is unsuited for the precise formulation of scientific theories. Also, we see different scientific disciplines all using their own specialized languages. These specialized languages distinguish themselves from the natural languages by their greater conceptual precision and, more often than not, by their significant curtailment of the colloquial field to which they specifically relate.

In general, several distinct aspects of a language can be distinguished (for similar distinctions, cf. Ch. Morris):

1. syntax, i.e., the grammar,
2. logic, i.e., the theory of conclusive reasoning (theory of inference),
3. semantics (also called 'theory of reference' (Quine)); Tarski describes this aspect as follows: "We shall understand by semantics the totality of considerations concerning those concepts which, roughly speaking, express certain connections between the expressions of a language and the objects and states of affairs referred to by these expressions" (Tarski (1), p. 401),
4. pragmatics, i.e., the use made of the language.

In a natural language, all these aspects are interwoven into a very complex coherence. In a specialized language, such as a mathematical language, all these distinct aspects are still present. What distinguishes a mathematical language from the natural language is not only the greater precision of the former but also the fact that it arises by eliminating much of what is mathematically irrelevant from the colloquial vocabulary. Also, quite often, one tries to eliminate one or more of the distinguished aspects. In this manner, one distinguishes, within mathematics, *pure mathematics,* by ignoring the pragmatic (applicative) aspect from mathematics as a whole. Further, one could also exclude the semantical aspect of a mathematical language, thereby obtaining a so-called *formal language.*

Assuming that all that which is commonly denoted by the term 'mathematics' could be formulated within a formal language, then one could truly say that mathematics is a purely formal science, and that, as a consequence, mathematical truth is nothing other than 'formal derivability from some set of axioms'. Such essentially formalistic assumptions suppress the semantic and pragmatic aspects of the 'full' mathematical language, not just for research purposes or for renovation of the mathematical activity, but to *reduce* mathematics to its formal aspects. Formalism, in this (philosophical) sense, hardly exists any more on account of the difficulties it has generated (cf. sections I.5, I.6 and II.4.3). Thus, if *we* delete the semantic and pragmatic aspects of mathematics, it is only for the sake of attaining a clearer insight into the formal side of a mathematical language. To delete semantics from a mathematical theory is to create a formal language without a truth concept, for in a formal language one cannot say whether or not something is true. The task of logic is not to determine what is true. Because, being only a 'theory of inference', it explicates the ways in which one reasons conclusively from given sentences, irrespective of their truth or falsity (e.g., modus ponens, modus tollens, etc.). The old philosophical formalism reduced the (semantical) demand for the truth of a mathematical theory to the (logical) demand for the intrinsic consistency of that theory, i.e., the theory may not engender contradictory conclusions.

In this section we shall introduce a formal language, and then, in the second and third sections, add the semantical aspect to it. The resulting language, containing syntactical, logical and semantical aspects, will, for the sake of simplicity, be called a *logical language*.

Before executing this program, we shall show, by means of two examples, that there are essential differences between natural language, formal language and logical language by mentioning the results of N. Chomsky and A. Tarski.

1. The structure of a formal language can often be program-

med for a computer (cf. Cohen, p. 3). The linguist Noam Chomsky, in his *Syntactic Structures,* investigates a number of models for certain grammars, the first of which is the so-called 'finite state grammar'. In 1949, Shannon and Weaver presented this grammar as a model capable of generating the sentences of a natural language, say, the English language. The model is based on a stochastic process, and produces sentences (Markov chains) in the following way: after having freely chosen a first symbol, the possible choices for the symbols that follow are always functions of the initial symbol(s) and their order.

A grammar which generates sentences according to this model can be programmed for a computer. But now Chomsky proves that such a 'finite state grammar' is incapable of even generating all the grammatically correct sentences of a colloquial language (such as English). He argues this from the fact that, in a 'finite state grammar', only a finite number of forms can be produced on account of the necessarily linear way in which the sentences are constructed. He says: "If grammar of this type produces all English sentences, it will produce many non-sentences as well. If it produces only English sentences, we can be sure that there will be an infinite number of true sentences, false sentences, reasonable questions, etc., which it simply will not produce" (Chomsky (1), p.24; cf. also Nivette).

Other types of models for the generation of a natural language seem to fail in a similar manner as well. Generalizing, we could say that the colloquial language is so 'rich' as to be incapable of being produced mechanically. This insight Chomsky also shares when he points to the essential qualitative chasm between the human capacity for language and that which an automaton could produce (Chomsky (2), p.6). In this book, we do not further pursue Chomsky's researches.

2. In conjunction with Tarski's concept of 'semantically closed languages', we are able to point out an essential difference between natural language and logical language. Every natural language has the property that one can discuss within it, the meaning of its terms and sentences, i.e., a natural language permits us to formulate or express its semantics in its own terms. Such a language, which contains its own semantics, Tarski calls 'semantically closed' (in Tarski (3): 'semantically universal'). He proves for such languages that, if they can be formalized, then they will be inconsistent (cf. Tarski (2); also see section I.7). (Here the possibility of formalization must be required because consistency, in an all-inclusive, imprecise colloquial language, is not a well-defined concept.) We shall expound the Tarski result later in this chapter.

Within a logical language one can distinguish several 'sublanguages', such as the language of the *propositional calculus,* that of the *predicate calculus,* etc. We shall now introduce the syntax (grammar) for the elementary predicate (or class) calculus, of which the propositional calculus is a part.

DEFINITION 1. The language of the *elementary predicate calculus* includes the following symbols:

(1) propositional or sentential connectives: $\sim$, &, $\vee$, $\rightarrow$, $\leftrightarrow$
(2) quantifiers : $\exists$, $\forall$
(3) parentheses : (,)
(4) variables : $x, y, z, \ldots$
(5) predicates : $A, B, C, \ldots$.

In principle, it would be sufficient in a formal language merely to introduce these symbols. However, to elucidate our discussion about these symbols and to show the relation between formal language and natural language, we shall express the meaning of these symbols. '~' (*negation*) and '&'(*conjunction*) mean 'not' and 'and', respectively. The symbol 'v' (*disjunction*) is best described with 'or'. However, in natural language, the word 'or' is used in two distinct senses, namely, an exclusive and a non-exclusive sense. Logic uses the latter. The Latin word 'vel' has approximately the sense of 'or' in the non-exclusive sense, and we agree, once and for all, to use the sign 'v' for the non-exclusive disjunction. The sign '→' (called *implication*) may be described in natural language as 'if ... then ...'. In the natural language we usually mean thereby a formal implication, while logic uses it in the sense of a material implication. (The distinction between these two forms of implication will become evident after we have introduced the 'truth tables' of elementary predicate calculus.)The symbol '↔' is the sign for a *bi-conditional*, i.e., a two-way implication which can be described in natural language with 'if and only if ...' (cf. Suppes, Chapter 1, for a more explicit description of all this).

These connectives and the parentheses, together with symbols, say P, Q, ..., which denote propositions or judgments, form the set of symbols for the *propositional* or sentential *logic*. (In section I.2, we shall introduce the semantics for the propositional logic.)

The quantifiers '∀' and '∃' mean 'for all' and 'there exist(s)', respectively. Parentheses are introduced to guarantee unique readability of expressions without danger of confusion.

We now show how, by means of these symbols, formulas can be formed. By means of a recursive definition, a small number of precise rules will suffice to define a well-formed formula (abbrev. wff).

(Since, in these notes, nothing more than unique readability will be striven for, the use of parentheses may be suppressed somewhat).

DEFINITION 2. The following rules determine when a string of symbols forms a wff in the elementary predicate calculus.

Rule 1. If A is a predicate symbol and t is a variable, then $A(t)$ is a wff.

Rule 2. If U and V are wff's then so are $\sim U$, $U \,\&\, V$, $U \vee V$, $U \rightarrow V$ and $U \leftrightarrow V$.

Rule 3. If U is a wff, x is a variable, then $(\forall x)U$ and $(\exists x)U$ are wff's.

Note that, in the *elementary* predicate calculus a predicate A can be a function of at most one variable, i.e., A is of the form $A(t)$. In the predicate calculus (Section I.5), a predicate A may be a function of more than one variable, say $A(x, y, z)$. This means, among other things, that in the predicate calculus, relations, such as $x < y$, x and y variables, are expressible, whereas in elementary predicate calculus they are not. An example of a wff in elementary predicate calculus is:

$$(\forall x)\,(\exists y)\,(A(x) \rightarrow (B(y) \vee ((\exists z)C(z)))).$$

For, according to Rule 1,

$$A(x), B(y) \text{ and } C(z)$$

are wff's; furthermore,

$$(\exists z)C(z)$$

is a wff according to Rule 3; then

$$B(y)\vee((\exists z)C(z))$$

is a wff by Rule 2, as is also

$$A(x)\rightarrow(B(y)\vee((\exists z)C(z))).$$

Finally, Rule 3 guarantees that

$$(\exists y)\,(A(x)\rightarrow(B(y)\vee((\exists z)C(z))))$$

and

$$(\forall x)\,(\exists y)\,(A(x)\rightarrow(B(y)\vee((\exists z)C(z))))$$

are wff's.

Note that, if P and Q are propositions, then a formula of the form

$$((P\rightarrow Q)\vee(Q\vee P))\rightarrow(P\rightarrow Q)$$

is a formula of the propositional calculus, whereas a formula of the form

$$(\forall x)\,(P\rightarrow Q)$$

is not.

Furthermore, note that Rule 3 allows expressions such as $(\forall x)\ A(y)$ as a wff. The latter expression is to be interpreted intuitively as meaning that if $(\forall x)$ or $(\exists x)$ occurs before a wff A, which does not involve x, then the effect of the quantifier is nil, and it can be omitted. This brings us to the precise distinction between *bound* and free *variables*. The occurrence of a variable in a formula can be one of two types: either that occurrence is controlled by a quantifier, or it is not. On the basis of Definition 2, this distinction will now be defined.

DEFINITION 3. Each occurrence of a variable symbol in a wff is defined as free or bound as follows:

(1) every variable occurring in a formula of the form mentioned in Rule 1 is free,

(2) the free and bound occurrences of variables in the wff's mentioned in Rule 2 are precisely the same as those for U and V separately,

(3) the free and bound occurrences of a variable in a formula $(\forall x)U$ or $(\exists x)U$ are the same as those for U except that every free occurrence of x is now considered bound.

DEFINITION 4. A *statement* is a wff with no free variables.

A mathematical theory may be thought of as a set of statements joined by rules of deduction.

I.2. SEMANTICAL INTERPRETATION OF THE PROPOSITIONAL CALCULUS

In the first section, we found that the propositional calculus forms a part of the elementary predicate calculus. The propositional calculus is concerned only with statements and their composition via connectives (Definition 1 in Section I.1). Logic treats its statements and propositions, whether true or false, independently of the question whether or not a decision procedure exists which could determine whether a particular proposition is indeed true or false. This 'truth-value' of the complex proposition in the propositional logic is determined by the 'truth-value' of the propositions composing the complex proposition. In (the area of) semantics, one is able to determine the truth-value of a complex proposition via the so-called 'truth-tables'.(For an approach to truth tables or propositional function tables which is more formal and more general, cf. Cohen, p. 8ff., and Tarski (4), p. 41ff., and Section I.3).

The truth-functional rules of usage for negation, conjunction, disjunction, implication and bi-conditional may be summarized in tabular form. These basic truth tables tell us at a glance under what circumstances the negation of a sentence is true if we know the truth (denoted by T) or falsity (F) of the sentence; similarly for the conjunction of two sentences, and the disjunction or implication of two sentences as well.

Negation

P	$\sim P$
T	F
F	T

Conjunction

P	Q	$P \,\&\, Q$
T	T	T
T	F	F
F	T	F
F	F	F

Disjunction

P	Q	$P \vee Q$
T	T	T
T	F	T
F	T	T
F	F	F

Implication

P	Q	$P \rightarrow Q$
T	T	T
T	F	F
F	T	T
F	F	T

Bi-conditional

P	Q	$P \leftrightarrow Q$
T	T	T
T	F	F
F	T	F
F	F	T

From these 'basic truth tables' one can, in a recursive manner, derive the truth tables for the complex propositions, e.g., of

$$(((\sim P) \vee Q) \rightarrow R),$$

as follows,

P	Q	R	$\sim P$	$((\sim P) \vee Q)$	$(((\sim P(\vee Q) \rightarrow R)$
T	T	T	F	T	T
F	T	T	T	T	T
T	F	T	F	F	T
F	F	T	T	T	T
T	T	F	F	T	F
F	T	F	T	T	F
T	F	F	F	F	T
F	F	F	T	T	F

A proposition which is always true, no matter what the truth values of its proposition letters may be, is called a *tautology*. A proposition which is false for all possible truth values of its proposition letters is called a *contradiction* (cf. also Suppes, Mendelson).

The (*formal*) *implication*, which is used more in colloquial language, implicitly assumes, in contradistinction to *material implication*, that the antecedent in the implication is always true. Consequently, its 'truth table' should be the following:

P	Q	if P then Q
T	T	T
T	F	F

In general, it can be said that there is no fixed way in which the natural language uses the implication, because the intrinsic intertwinement of the semantics of natural language expressions with the immediate context, does not necessitate it (cf. also Beth (1), Tarski (3)).

I.3. SEMANTICAL INTERPRETATION OF THE ELEMENTARY PREDICATE CALCULUS.

In Section I.1, we formulated the syntax of the elementary predicate calculus, i.e., the symbols were given, with the rules for connecting them. Neither the logical aspect (the rules of deduction) nor the semantical aspect of this formal language were brought to expression. In Section I.2. we merely introduced the (semantic) 'truth rules' for the propositional calculus. In this section we shall deal with the semantics of the broader elementary predicate calculus. To that end, we begin with the *semantical interpretation* of a predicate- (or class-) logical wff U.

We may fix (the meaning of) a certain predicate or a predicate symbol in the same way as a variable can be fixed in a 'universe of discourse', e.g., in the set of all people, or the set of integral numbers. In this manner, the formula

$$(\forall x)\,(A(x) \rightarrow B(x))$$

could mean: every number greater than 3 is greater than 2. In such cases, we speak of an *interpretation* of a formula.

DEFINITION 1. If the truth range of the variables of a formula U is S, and if a, b, c, ... are properties denoted by A, B, C, ..., then we call the system $[S; a, b, c, ...]$ an *interpretation of the formula U.*

It is clear that one formula can have many interpretations. The above formula

$$(\forall x)(A(x) \rightarrow B(x))$$

could also be interpreted in the set of all mammals, i.e., in another universe of discourse, where A could mean 'is a tiger' and B 'has only one leg'. This example shows that not every interpretation of a formula must be true.

A *propositional function* is a formula which does not contain quantifiers, i.e., no '∀' or '∃'. To each formula we now want to add a *truth value*, depending on a given interpretation. We do this by induction on the length of the formulas.

DEFINITION 2. Let U be a formula with free variables among $x_1, \ldots, x_n$, $(n > 0)$ and let $\bar{x}_1, \ldots, \bar{x}_n$ be elements of the universe S of the interpretation. We define the *truth value* of U (in S) at $\bar{x}_1, \ldots, \bar{x}_n$ as follows:

(1) if U is $A(x)$, where A is a predicate symbol and x is one of the $x_1, \ldots, x_n$, then U is true at $\bar{x}_1, \ldots, \bar{x}_n$ if $\bar{x}$ (one of the $\bar{x}_1, \ldots, \bar{x}_n$) has the property a, i.e., $a(\bar{x})$;

(2) if U is a propositional function of formulas, the truth of U at $\bar{x}_1, \ldots, \bar{x}_n$ is established by means of the truth tables for the propositional calculus;

(3) if U is of the form $(\forall y) V(y, x_1, \ldots, x_n)$ [resp. $(\exists y) V(y, x_1, \ldots, x_n)$] then U is true at $\bar{x}_1, \ldots, \bar{x}_n$ if, for all $\bar{y}$ in S [resp. for some $\bar{y}$ in S], the formula $V(y, x_1, \ldots, x_n)$ is true at $\bar{y}, \bar{x}_1, \bar{x}_2, \ldots, \bar{x}_n$.

We note that if V is a statement, we can take $n = 0$, and our definition is just truth in S under the given interpretation. A wff with free variables stands for a *relation* on the universe of discourse of the interpretation, and may thus be satisfied (true) at some 'place' in the universe but not satisfied (false) at another 'place'.

DEFINITION 3. We call an interpretation $[S; a, b, c, \ldots]$ of a statement U a *model* for U, if U is true when interpreted thus.

In general, not every interpretation of a statement will form a model for U. We saw this in a previous (false) interpretation of the formula

$$(\forall x)\,(A(x) \to B(x)).$$

DEFINITION 4.

(a) If every interpretation $[S; a, b, c, \ldots]$ of U is a model for U, then U is called a *logical identity* or a *tautology*.

(b) If no interpretation of U is a model for U, then U is called a *logical contradiction*.

(c) If U is neither a logical contradiction nor a logical identity, then U is called *logically neutral*.

As examples of tautologies, we give:

$$(\forall x)\,(A(x) \to A(x));$$

$$(\forall x)\,(A(x) \vee \sim A(x)),$$

$$(\forall x)(A(x) \,\&\, B(x) \,\&\, C(x)) \vee (\exists x)(\sim A(x) \vee \sim B(x) \vee \sim C(x)).$$

The last statement is a tautology because for every interpretation of it, it is true. (We assume that our universe is not empty).

The intuitive notion undergirding logical identity is that an identity will be true in all 'possible worlds' (cf. Section II.2 and III.3 on Leibniz).

We can also formulate weaker identity and contradiction concepts by limiting all 'possible worlds' to a certain 'universe of discourse' S. We then speak of an S-identity, S-contradiction and S-neutral.

DEFINITION 5. If every interpretation $[S; a, b, c, \ldots]$ of a statement U, with a fixed universe of discourse S, is a model for U, then U is an *S-identity*.

Analogous definitions can be stated for the concepts S-contradiction and S-neutral.

From the above definitions, the following theorems can be derived.

(i) U is an identity if U is an S-identity for every choice of the universe S.

(ii) U is a contradiction if U is an S-contradiction for every choice of the universe S.

(iii) If, for at least one choice of S, U is S-neutral, then U is neutral.

The proofs of these theorems are easily given, as the reader can himself readily verify.

I.4. DECISION PROCEDURE FOR THE ELEMENTARY PREDICATE CALCULUS

A procedure, which enables us to determine whether a formula is a logical identity, logical contradiction or is logically neutral, is called a *decision procedure*. In this section we will show that such a procedure exists for the elementary predicate calculus. This procedure was first indicated by Behmann (1922).

THEOREM 1. *Let* $[S; a, b, c, \ldots]$ *be a model for* U *and let* S' *be a set equipotent with* S, *then we can choose properties* $a', b', c', \ldots$ *in* S' *in such a manner that* $[S', a', b', c', \ldots]$ *is also a model for* U.

PROOF. Because S is equipotent with S', there is a 1-1 correspondence between the elements of S and S'. If we take a particular 1-1 correspondence, then each element q of S' is correlated which an element p of S. The latter element we denote as $p(q)$. We now construe the predicate a' in such a manner that the element q of S' has this property a' only if the element $p(q)$ of S, correlated with q, has the property a. In an analogous fashion,construe the properties $b', c', \ldots$ for the elements of S'. The interpretation $[S'; a', b', c', \ldots]$, constructed in this fashion, is then a model for U which is 'homomorphic' to the model $[S, a, b, c, \ldots]$ for U.

With the help of, among other things, Theorem 1, it is easy to prove that a statement remains an S-identity, S-contradiction or S-neutral for every interpretation with respect to a universe of discourse which is equipotent with S; i.e., the question whether a formula is an identity, contradiction, or is neutral, is not determined by the nature of the elements of a universe, but only by the 'extension' of the universe.

It is now possible to obtain a real decision procedure whereby, by means of an investigation of one finite model for a formula, it can be decided whether that formula is an identity, a contradiction, or is neutral.

In the following theorem, by $U[A, B, C, D]$ we shall mean a statement which contains, as predicate symbols, only A, B, C and D.

THEOREM 2a. *Let* $[S; a, b, c, d]$ *be a model for* $U[A, B, C, D]$. *Then there is also a model* $[S'; a, b, c, d]$ *for* U *whereby* S' *contains no more than* $2^4 = 16$ *elements.*

PROOF. We give only an outline of the proof. Let $[S; a, b, c, d]$ be a model for U. Then the elements of S can, at most, form 16 classes. Each element of S belongs to only one of the following classes:

Class 1 - its elements have the properties a, b, c and d;
Class 2 - its elements have the properties a, b, c not d;
⋮
Class 16 - its elements do not have the properties a, b, c, or d.

For each class, the truth of U is unambiguously and uniquely determined by the interpretation $[S; a, b, c, d]$.

It may happen that in certain cases the classes are empty. Such would be the case if each element of S had the property d, i.e., there is no element of S that has the property not-d. For then at least the Classes 2 and 16 above would be empty. (For this reason we say 'no more than' in the formulation of the theorem). Now select, as representative of each non-empty class, one element that belongs to each respective class. In this manner, we can form a set S', with no more than 16 elements. It is clear that $[S'; a, b, c, d]$ is now an interpretation of $U[A, B, C, D]$. Since $[S; a, b, c, d]$ is a model for U, so also $[S'; a, b, c, d]$ (because of the defined homomorphism that leaves the truth intact).

We can also formulate this theorem in a more general form:

THEOREM 2b. *Let* $[S; a_1, \ldots, a_n]$ *be a model for the statement* $U[A_1, \ldots, A_n]$, *where* $A_1, \ldots, A_n$ *are predicate symbols. Then there is also a model* $[S'; a_1, \ldots, a_n]$ *for* U *with* $S' \subset S$ *and with cardinality* $\leqslant 2^n$.

This theorem gives us immediately an important result. For it now follows that number theory cannot be formulated in the elementary class logic. Were this the case, then for each number-theoretical theorem there would be a finite model, and hence a decision procedure would exist in all cases. But this seems not to be the case: e.g., the unresolved problem of Fermat: find integers x, y and z which satisfy.

$$x^n + y^n = z^n, \text{ for } n > 3. \text{ (cf. I.5.)}$$

Note. An arbitrary subset of the universe of a model need not necessarily generate a new model. E.g.,

$$((\exists x)A(x) \; \& \; ((\exists y) \sim A(y))$$

has a model, $[\{0, \pm 1, \pm 2, \ldots\}; < 0]$; the couple $[\{0, \pm 1\}; < 0]$ is a model as well. But the couple $[\{0, 1\}; < 0]$ apparently fails to be a model.

THEOREM 3. *If the interpretation* $[S; a, b, c, d]$ *is not a model for* U, *then it is a model for* $\sim U$.

PROOF. This follows directly from the definition of negation and the fact that the interpretation of a statement must be either true or false.

THEOREM 4. *A statement U with 4 predicate symbols is an identity (contradiction, resp.) if and only if it is an S-identity (S-contradiction, resp.) for a universe of no more than 16 elements.*

PROOF. Make use of Theorems 1 and 2.

Generally, it is not for one formula that we want a model but for a whole theory (i.e., a set of statements). To enable us to formulate this notion we give a few more theorems and definitions.

We can expand the notion of the interpretation by *interpreting a set of statements*, i.e., by interpreting all the predicates in the formulas in one universe of discourse.

DEFINITION 1. $[S; a, b, c, \ldots]$ is a *model* for a set of statements α, if each statement of α is true in $[S; a, b, c, \ldots]$.

DEFINITION 2. A statement W is a *conclusion* from statements U and V, if each common model for U and V is also a model for W.

THEOREM 5. *A statement W is a conclusion from statements U and V, if, and only if, the statement $(U \& V) \rightarrow W$ is an identity.*

PROOF. (i) Assume $(U \& V) \rightarrow W$ is not an identity. Then there is an interpretation $[S; a, b, c, \ldots]$ which is not a model for $(U \& V) \rightarrow W$; i.e., $(U \& V) \rightarrow W$ interpreted in this manner is false; i.e. $(U \& V)$, and hence U and V, is (are) true in this interpretation but W is false (according to the truth definition of the implication). But then $[S; a, b, c, \ldots]$ is a model for U and V, but not for W. Consequently, by definition, W is not a conclusion from U and V.

(ii) Assume that $(U \& V) \rightarrow W$ is an identity and that $[S; a, b, c, \ldots]$ is a model for U and V. Then it is also a model for $(U \& V) \rightarrow W$ (the model also interprets the predicates of W!) according to the definition of logical identity. If $(U \& V) \rightarrow W$ and $(U \& V)$, when both interpreted in $[S; a, b, c, \ldots]$, are true, then W is also true. This means that $[S; a, b, c, \ldots]$ is a model for W.

THEOREM 6. *The statement W is not a conclusion from U and V, if U, V and $\sim W$ possess a common model.*

PROOF. Easy.

Remark. The logic of Aristotle encompasses the so-called modal logic; i.e., a logic which includes judgments in which terms such as probable, ought to, necessary, possible, etc. appear. In the history of the subject since Aristotle, much critique has been directed to this part of his logic. It is not surprising then that Aristotle's 'assertoric logic' has received all the attention. This assertoric logic can easily be formulated within the elementary predicate calculus. For the standard form of the Aristotelian judgment is the S-P (subject-predicate) scheme (e.g., 'all things are extended'), which, in our logical language, can be formulated as $A(x)$: x has the property A. There remain, however, essential differences between the classical syllogistic forms of reasoning and the elementary predicate calculus, which is especially geared to mathematical theories.

In this manner, for example, the traditional syllogistic form of reasoning known by the name Barbara is also valid in the elementary predicate calculus:

All men are mortal
Socrates is a man
Socrates is mortal

In the elementary predicate calculus this is formulated as:

$$\begin{array}{ll} U: & (\forall x)(A(x) \rightarrow B(x)) \\ V: & (\forall x)(B(x) \rightarrow C(x)) \\ \hline W: & (\forall x)(A(x) \rightarrow C(x)) \end{array}$$

and indeed, W is a conclusion from U and V.

It turns out, however, that the syllogistic forms known as Felapton and Fesapo are not, in our terminology, conclusive forms of reasoning. This latter form is:

$$\begin{array}{ll} U: & (\forall x)(B(x) \rightarrow A(x)) \\ V: & (\forall x)(B(x) \rightarrow \sim C(x)) \\ \hline W: & (\exists x)(A(x) \rightarrow \sim C(x)) \end{array}$$

That W is not a conclusion from U and V follows from the definition (of the truth table) of material implication. Modern logic uses material and not formal implication (cf. Section I.2).

DEFINITION 3. A set of sentences $U_1, \ldots, U_n$ is called *non-contradictory* when is does not contain two contradictory conclusions.

THEOREM 7. *The sentences $U_1, \ldots, U_n$ are non-contradictory, if and only if, they possess a common model.*

PROOF. Easy.

The totality of the theorems treated so far enables us to

solve the so-called 'Entscheidungsproblem' for elementary predicate calculus, namely: *we have available an unfailing procedure which enables us to determine, in a finite number of steps, whether:*

(1) *a given statement is an identity,*

(2) *a given statement is a conclusion from statements* $U_1, \ldots, U_n$,

(3) *a given set of statements* $U_1, \ldots, U_n$ *is contradictory.*

This can be readily seen. For, case (1) is determined with the help of Theorem 4; case (2) can be reduced to case (1) by asking whether $(U_1 \ \& \ \ldots \ \& \ U_n) \to U$ is an identity (Theorem 5); and similarly for case (3) by asking whether $\sim (U_1 \ \& \ \ldots \ \& \ U_n)$ is an identity (Theorem 7).

We mentioned already that the elementary predicate calculus cannot provide a sufficient basis for number theory. It appears that we need a stronger language to formulate, for instance, the problem of Fermat. Such a language will now be introduced.

I.5. PREDICATE CALCULUS – THE THEORY *Z*.

In elementary predicate calculus we cannot formulate the problem of Fermat because, for its formulation, we need a relation symbol. In the elementary predicate calculus we meet predicates with only one variable, or single-valued predicates. In the predicate calculus we also operate with many-valued predicates, or relations. This means that number-theoretical relations such as $x < y$, x/y, etc. (x and y are variables) can now be formulated. (Please note that the 'relation' $x < 0$, 0 a constant, in the previous section, we took to be a single-valued predicate). The set of symbols in predicate calculus will be an expansion of that in elementary predicate calculus (cf. Section I.1, Definition 1).

DEFINITION 1. The language of the *predicate calculus* contains the following symbols:

(1) connectives, quantifiers, parentheses and variables as in the language of elementary predicate calculus,

(2) relation symbols $A_1, A_2, \ldots$ whereby to each A_i is assigned an integer $n_i \geqslant 1$, n_i indicating that the relation represented by A_i is a relation between n_i objects (n_i-ary relation symbol),

(3) the symbol of equality: '=',

(4) constants: $c_1, c_2, c_3, \ldots$.

DEFINITION 2. Rules for well-formed formulas:

Rule 1. $x = y$, $x = c_1$, $c_1 = c_2$ are wff's, where x and y are variable symbols and c_1 and c_2 are any constant symbols.

Rule 2. If A is an n-ary relation symbol and each of $t_1, \ldots, t_n$ is either a variable or constant symbol, then $A(t_1, \ldots, t_n)$ is a wff.

Rule 3. If U and V are wff's, then so are U, $U \,\&\, V$, $U \vee V$, $U \rightarrow V$ and $U \leftrightarrow V$.

Rule 4. If U is a wff, then so are $(\exists x)U$ and $(\forall x)U$.

As in elementary predicate calculus, we are able to introduce into predicate calculus the concepts free and bound variable, statement, interpretation and model. No essential changes need be made to incorporate them into predicate calculus.

The recursive definition of an interpretation must be expanded so that the constant symbol can also be interpreted in the universe of discourse. The interpretation of '=' will be that of identity. With respect to this expanded concept of interpretation, we can quite simply introduce the expanded concept of a model (cf. Cohen, pp. 12, 13).

Before proceeding further with the relation between the formal structure of predicate calculus and the models for it, we should mention first that the above rules for the formation of sentences and statements in predicate calculus do not suffice to formulate *all* of what is usually called 'elementary number theory'. As an example we have Fermat's last conjecture. It states that for all positive integers n ($n \geqslant 3$), the equation $x^n + y^n = z^n$ has no solution if x, y and z are integers. How can we express statements such as this in the formalism of predicate calculus? First, we remark that an arithmetical relation, such as $x \cdot y = z$, is, in fact, a 3-ary predicate, say $A(x, y, z)$. Furthermore, $a + b = c$ is also a 3-ary predicate, say $B(a, b, c)$. Thus, to express the fact that there are integers x, y and z satisfying the equation $x^2 + y^2 = z^2$, we must write:

$$(\exists x)\,(\exists y)\,(\exists z)\,(\exists a)\,(\exists b)\,(\exists c)\,(A(x, x, a)\ \&\ A(y, y, b)\ \&\ A(z, z, c)\ \&\ B(a, b, c)).$$

It is immediately clear that, in order to express the general Fermat statement, it is necessary to let a variable, say n, run through the number of free variables occurring in an n-ary predicate. Because such things are not permitted, we can only formulate Fermat's last conjecture in predicate calculus by considering it as an *infinite number* of statements. We shall see that the same sort of thing happens if we try to formulate the (number theoretic) axiom of induction in predicate calculus. The fact, however, that predicate calculus is nevertheless considerably 'stronger' than elementary predicate calculus is readily seen if one considers the fact that, in predicate calculus, statements occur which admit infinite models and no finite ones. Take, for instance, Schütte's statement:

$$(\forall x)(\sim A(x,x) \;\&\; (\exists y)(A(x,y) \;\&\; (\forall z)(A(z,x) \rightarrow A(z,y)))).$$

This statement says that:

(1) no element p of a model S (for this statement) has the relation A to itself (i.e., $\sim A(x, x)$);

(2) for every p of S, there exists a q of S such that p has the relation A to q;

(3) each r in S, which has the relation A to p, also has a relation A to q.

In this manner, we get an infinite sequence of elements in S, if S has at least one element.

In our colloquial language we have a notion of truth to which the concept of truth in the propositional calculus still strongly appeals. In the elementary predicate calculus we introduced the concept of truth by means of a model, and this was an expansion of the propositional calculus. All these definitions of truth are *semantic* definitions. It turns out that, in logic - the logical aspect of a formal language - there are notions whose extent strongly correspond to that of the semantical concept of truth. Logic is a 'theory of inference' we have said earlier. Hence our rules of inference for predicate calculus have to be such that only true formulas are derivable from true formulas. In this manner we come to the logical concept of a 'valid statement' (cf. Cohen, p. 8ff.), whereby we wish the set of 'true statements' (semantically determined) to be the same as those of 'valid statements' (logically determined); just as is the case with the rules of inference of the propositional calculus. It is therefore via the propositional calculus that we lay a connection between logical identities (tautologies) and valid statements. This is expressed in the first rule of inference of the predicate calculus. We have altogether seven rules of inference.

DEFINITION 3. Rules of inference of the predicate calculus.

Rule 1. If U is a tautology (in the sense of the propositional calculus) built up from the propositions $V_1, \ldots, V_n$, then the result of replacing each V_i by any statement of the predicate calculus is a *valid statement.*

Rule 2. (*Modus Ponens*) If U and $U \to V$ are valid statements, then so is V.

Rule 3. (*Rules of Equality*)

(1) $c_1 = c_1$, $(c_1 = c_2) \to (c_2 = c_1)$ and $((c_1 = c_2) \mathbin{\&} (c_2 = c_3)) \to (c_1 = c_3)$ are valid statements, where c_1, c_2 and c_3 are any three constant symbols,

(2) If U is a statement, c_1 and c_2 constant symbols, and if U' represents U with every occurrence of c_1 replaced by c_2, then $(c_1 = c_2) \to (U \to U')$ is a valid statement.

Rule 4. (*Change of Variables*) If U is any statement and U'

results from U by replacing each occurrence of the symbol x with the symbol x', where x and x' are any two variable symbols, then the statement $U \rightarrow U'$ is a valid statement.

For the next rule, let $U(x)$ represent a formula with one free variable x and in which *every* occurrence of x is free and let $U(c)$ represent the result of replacing every occurrence of x by the constant symbol c.

Rule 5. (*Rule of Specialization*) $((\forall x)U(x)) \rightarrow U(c)$ is a valid statement, where c is any constant symbol.

The next rule is a bit misleading and requires some explanation. Often in arguments we say "let c be an arbitrary but fixed integer". We then proceed to reason about c and come to a certain conclusion $A(c)$. We can then deduce that $(\forall x)A(x)$ since we use no special properties of c. In reality, we have treated c as a variable, even though we called it a constant. This is because our valid statements will be true in every interpretation of the constant and relation symbols. We could express this by: if $A(c)$ is a valid statement, then so is $(\forall x)A(x)$. However, Rule 6 is in a more convenient form and yields the same results.

Rule 6. Let V be a statement not involving c of x. Then if $U(c) \rightarrow V$ is valid, then so is $(\exists x)U(x) \rightarrow V$.

Rule 7. Let $U(x)$ have x as the only free variable and let every occurrence of x be free. Let V be a statement which does not contain x. Then the following are valid statements:

$$(\sim (\forall x)U(x)) \leftrightarrow ((\exists x) \sim U(x))$$

$$(((\forall x)\, U(x))\ \&\ V) \leftrightarrow ((\forall x)\, U(x)\ \&\ V)$$

$$(((\exists x)\, U(x))\ \&\ V) \leftrightarrow ((\exists x)\, U(x)\ \&\ V).$$

DEFINITION 4. Let α be a collection of statements. We say that U is *derivable* from α, if for some $V_1, \ldots, V_n$ in α, the statement $(V_1\ \&\ \ldots\ \&\ V_n) \rightarrow$ U is valid.

Exercise. Prove that $U(c)$ is valid, if $(\forall x)\ U(x)$ is valid.

Having now given rules for forming valid statements we come to the problem of identifying these statements with the true statements.

DEFINITION 5. A set of statements α is said to be *consistent* if the statement $(U\ \&\ \sim U)$ cannot be derived from α for any U.

The correctness of the following theorem is immediately evident when one verifies the theorem for the different rules of the predicate calculus (tedious).

THEOREM 1a. *If U is a valid statement, then it is true in every model.*

Hence, the set of logical identities is a subset of the set of valid statements.

THEOREM 1b. *If a set of statements* α *has a model, then it is consistent.*

PROOF. Assume that α has a model, but is not consistent. Then there is a formula U in α such that $(U\ \&\ \sim U)$ is true in that model. But this means that U and $\sim U$ are both true in that model, which is impossible.

The converse of this theorem holds also.

THEOREM 2. (*Gödel's completeness theorem*). *Let α be any consistent set of statements. Then there exists a model for α whose cardinality does not exceed the cardinality of the number of statements in α if α is infinite, and is countable if α is finite.*

We will not give the complete proof of this theorem (cf. Cohen, or for a more explicit account, Kleene). The proof makes use of the axiom of choice and the elementary theory of sets. The proof consists of a method of making the required model. The proof however is non-constructive in the sense that the construction of the model for α may depend upon examining an infinite number of possibilities. The first step shall be the proof of the completeness of a subsystem of the predicate calculus.

THEOREM 3. (*Completeness of the Propositional Calculus*). *If α contains no quantifiers and is consistent, then there is a model for α.*

The model constructed to prove Theorem 3 can be expanded so as to serve as a model that proves the completeness theorem of Gödel (Theorem 2). We shall only state and prove one lemma that is used in the proof of Theorem 3.

LEMMA. *If α is a consistent set of statements, U an arbitrary statement, then either $\alpha \cup \{U\}$ or $\alpha \{\sim U\}$ is consistent.*

PROOF. If $\alpha \cup \{U\}$ is inconsistent, then, for some V_i in α, $(U \ \& \ V_1 \ \& \ \ldots \ \& \ V_n) \rightarrow (W \ \& \sim W)$, for some W, *is valid. If* $\alpha \cup \{\sim U\}$ is inconsistent, then, for some V_i' in α, $((\sim U) \ \&$

$V_1' \ \& \ \dots \ \& \ V_m') \to (W \ \& \sim W)$ is valid. The propositional calculus now implies that $(V_1 \ \& \ \dots \ V_n \ \& \ V_1' \ \& \ \dots \ \& \ V_m') \to (W \ \& \sim W)$ is valid, so that α must be inconsistent.

We have now constructed a formal language and we now want to show how a mathematical theory, such as, for example, elementary arithmetic, can be formulated in it. The system of the predicate calculus becomes concrete in that the predicates A, B, C, ... now come to represent relations in elementary number theory; so also, the constants c_1, c_2, c_3, ... will represent definite elements, e.g., $c_1 = 0$, etc. In this manner, we formulate an elementary mathematical theory, the theory of numbers Z, simply by giving a number of axioms, which reflect the rules for operating with the natural numbers within predicate calculus.

We shall not completely formalize this system, i.e., introduce it in the notation of the predicate calculus. For we wish to remain close to what might be called intuitive arithmetic. The elements in Z are to be thought of as the non-negative integers. We introduce ternary relations R_1 and R_2 corresponding to addition and multiplication, respectively. To make our formulas more readable, we shall write $x + y = z$ and $x \cdot y = z$ in place of what would otherwise be $R_1(x, y, z)$ and $R_2(x, y, z)$. We use two constant symbols, 0 and 1. Also, we introduce the notation $(\exists! x)A(x)$ as an abbreviation for $((\exists x)\ (\forall y)A(y)) \leftrightarrow (x = y)$. (This shows that we have formalized the notion "there exists a unique x such that $A(x)$".).

DEFINITION 6. The axioms for Z are the following:

1. $(\forall x, y)\ (\exists! z)\ (x + y = z)$;

2. $(\forall x, y)\ (\exists! z)\ (x \cdot y = z)$;

3. $(\forall x)\ (x + 0 = x)\ \&\ (x \cdot 1 = x))$;

4. $(\forall x, y)((x + (y + 1)) = ((x + y) + 1))$;

5. $(\forall x, y)((x \cdot (y + 1)) = ((x \cdot y) + x))$;

6. $(\forall x, y)((x + 1 = y + 1) \rightarrow (x = y))$

7. $(\forall x)(\sim (x + 1 = 0))$.

Our eighth axiom consists actually of an infinite number of axioms and is more properly called an *axiom scheme*. To state it, we enumerate the many countable formulas of our system which contain at least one free variable, $U_m(x, t_1, \ldots, t_k)$, where, of course, k depends on m. The particular enumeration we choose is of no consequence.

$$8_m . (\forall t_1, \ldots, t_k)((U_m(0, t_1, \ldots, t_k) \mathbin{\&} (\forall y)(U_m(y, t_1, \ldots, t_k) \rightarrow U_m(y + 1, t_1, \ldots, t_k))) \rightarrow (\forall x)U_m(x, t_1, \ldots, t_k)).$$

This eighth axiom can be viewed as a formulation of the principle of mathematical induction.

These axioms form a possible formalization of the axioms of Peano, which are considered to be an adequate set of axioms for the elementary theory of the natural numbers.

I.6. GÖDEL'S INCOMPLETENESS THEOREM

While the completeness theorem showed that the rules of the predicate calculus are complete, the incompleteness theorem of Gödel proves that *within* each theory, which includes Z and is based on the predicate calculus, the decision problem (for the predicate calculus and the formulated formal theory in it) is unsolvable.

THEOREM 1. (*Gödel's first incompleteness theorem*). *If Z is consistent, then there exists within Z a statement such that neither the statement nor its negation is provable within Z.*

We shall only give a few schematic indications of the proof.

One could ask what is computable within the system of the natural numbers and what is not. In any case, all *primitive recursive functions*, i.e., functions which are defined on the integers (or *n*-tuples of integers) and take integral values, can be computed. The set of such functions can easily be defined by means of systems of equations (cf. Cohen, p. 27ff.; also, in more expanded form, Davis). Most elementary functions of integers are primitively recursive; for example, addition, multiplication and powers. The primitive recursive functions that are constructible form a subset of all the so-called *effectively computable functions*. 'Church's thesis' is that the class of effectively computable functions is the same as the so-called *general recursive* functions. This thesis can not be proven mathematically because the concept 'effectively computable' is not a mathematical one. There are, however, a number of arguments that can be forwarded in support of it; for instance, it turns out that all explications that mathematicians have given of the concept 'computable' are equivalent to the mathematically defined notion 'recursiveness'.

Gödel's definition (1934) is essentially that a function is general recursive if there is some finite set of equations involving the function and auxiliary functions, which define the function uniquely. Türing equated 'general recursiveness' with 'computable in a Türing machine', i.e., an idealized (in the sense that the memory space is always capable of expansion) calculating machine.

To prove the theorem one proceeds by the following method (diagonal method):

(1) The first concern is the arithmetization of Z. By this we mean that we assume the formulas of Z to be enumerated in a natural manner (e.g., by counting them). Once this is done, the various elementary operations such as forming negation and conjunction of formulas, bounding a free variable, etc. become rather simple primitive recursive functions on the integers corresponding to the formulas. These integers are called the *Gödel numbers* of the formulas. Now the rules of the predicate calculus are given quite effectively, although there are infinitely many axioms of Z. For they are generated by a simple rule and one can easily enumerate their Gödel numbers. This means that by enumerating all possible proofs, we see that the *provable statements* are the range of a primitive recursive function. We can say that we have *interpreted* the syntax of Z, denoted Syn(Z), in Z.

(2) We now enumerate the primitive recursive functions by counting the systems of equations that define them. We can construct a primitive recursive function F whose range is not recursive. More precisely, F will have the property that if f_r is the primitive recursive function defined by the rth set of equations, then $r + 1$ is in the range of $F \leftrightarrow f_r(r + 1) = 0$. The function $F(x)$ can be represented in Z in the sense that there is a formula in Z which says $F(x) = y$. Let V_n denote the statement 'n is in the range of F'. The Gödel numbers of V_n and $\sim V_n$ are primitive recursive functions of n.

(3) For each n, we go through the list of ordered provable statements until we encounter a proof of V_n of or $\sim V_n$. If we do, the calculation ceases and we put $g(n) = 1$ in the first case, $g(n) = 0$ in the second. If neither V_n nor $\sim V_n$ is provable then $g(n)$ remains undefined. The function g, which has thus been very explicitly defined, must be a f_r for a particular r

(which can be explicitly found). Let U be the statement '$r + 1$ is in the range of F'.

(4) F is so defined that $f_r(r + 1) \neq \chi(r + 1)$, where χ is the characteristic function of the range of F. I.e., $\chi(n) = 0$ or 1, if and only if, *n is not* in the range of F or *n is* in the range of F, respectively.

(5) Now the following holds: if the proof of U occurs before a proof of $\sim U$, then $r + 1$ is not in the range of F; and, if the opposite occurs, then $r + 1$ is in the range of F.

(6) Now a statement which is provable in Z is of course true in the model of the integers. If we assume that Z is consistent, then not both U and $\sim U$ are provable. Thus, if U is provable, the assertion implies $r + 1$ is not in the range of F (by (5) above). But if U is provable, then U is true and $r + 1$ is in the range of F; a contradiction. If $\sim U$ is provable, then by (5), $r + 1$ is in the range of F. But if $\sim U$ is provable, then $\sim U$ is true and $r + 1$ is not in the range of F; again a contradiction. Thus, neither U nor $\sim U$ is provable, and our theorem is proved.

COROLLARY. U is false in the integers.

PROOF. If $r + 1$ were in the range of F, then for some m, $F(m) = r + 1$. Since F is recursive, we can write down all the steps of the calculation of $F(m)$. These calculations would then constitute a proof in Z of $F(m) = r + 1$ and hence a proof of U. But U is unprovable in Z. (Proof informal)

To sum up: on the one hand $\sim U$ is not provable in Z and yet we have just given an informal proof that $\sim U$ is true. This suggests, as the only remaining conclusion, that our informal reasoning uses a *natural principle*, which cannot be formalized in Z. In other words, *the system Z is so poor syntactically, that it cannot encompass our natural way of reasoning*. For a popularized account of Gödel's result consult also Nagel & Newman.

The method used by Gödel in his proof is essentially Cantor's diagonal method. With it, Cantor proved that uncountable sets existed. He did this by showing the necessary existence of an element, that was clearly a member of the set of real numbers on the continuum (the unit interval (0,1)) but could not be counted in a possible enumeration of the elements of this set in decimal notation (cf. Kamke, p. 12 ff).

Chomsky uses a similar method when he points out that the finite state grammar (cf. Section I.1) cannot generate all the sentences of a natural language: "the point is that there are processes of sentence formation that finite grammars are intrinsically not equipped to handle" (Chomsky (1), p. 23).

The incompleteness theorem of Gödel carries many consequences. For one thing, it meant the end of all speculation about the Türing machine as a machine that thinks, or as a machine by which one could adequately simulate mathematical thinking (cf. for this problem Beth (1), especially Chapter 5: 'Over de zogenaamde denkmachine'). For Gödel's theorem says that statements exist of which, within the system, neither the affirmation nor the denial can be proven. This means, that there are true formulas which cannot be deduced within *Z*. Thus a Türing machine will not be able to generate every true formula of a system (wherein *Z* is formulated).

Gödel's result also made an abrupt end to Hilbert's formalistic program. This program had as a goal, the proof of the consistency for the whole of mathematics, and made use of the notion that mathematical truth is equivalent to deducibility in some formal system (cf. Sections I.1 and II.4.3).

The end of the program followed immediately from Gödel's second incompleteness theorem (the proof of which follows in a non-trivial manner from that of his first theorem):

THEOREM 2. *The consistency of Z cannot be proven within Z.*

Formalism, after 1931 (the year in which Gödel published his results), has turned to the so-called *theory of models*. It has thus tried to fulfill its program in a weaker form, by giving *relative consistency proofs*. If one interprets a theory α within a theory β, and if this interpretation forms a model for α, then one has proven the consistency of α, presupposing of course that theory β is consistent (compare with Section I.5, Theorem 1b). In this manner, within the (3-dimensional) Euclidean geometry there exists a model for the hyperbolic geometry, namely, by interpreting the lines on a hemisphere as geodesics in hyperbolic geometry. If Euclidean geometry is consistent, then hyperbolic geometry is also consistent.

A similar kind of relative consistency proof is obtained as follows. Let $U_1, \ldots, U_n$ be the axioms of a certain mathematical theory μ. Assume that all but one of the axioms, say $U_1, \ldots, U_{n-1}$, form a logically consistent system. Then develop methods to show that the addition of the axiom U_n to the system $U_1, \ldots, U_{n-1}$, does not lead to a contradiction. Consistency proofs of this kind often proceed by constructing a model M for the axiomatic subtheory of μ, defined by $U_1, \ldots, U_{n-1}$, and then manipulate, interpret, or modify the model M in such a way that all the axioms $U_1, \ldots, U_n$ are satisfied in it. (For instance: the so-called Continuum Hypothesis of Set Theory is consistent with the usual axioms of Set Theory, and so are a number of its negations, cf. Cohen, Rosser (2).)

There is also the concept of the *independence of a certain axiom* from the other axioms of a theory μ. We say, using the above notation, that the axiom U_n is *independent* of the axioms $U_1, \ldots, U_{n-1}$ if there is a model for the system $U_1, \ldots, U_{n-1}$, in which U_n is not satisfied. Thus, for instance, in the classical Euclidean geometry, the 'axiom of parallels' is independent of the other axioms, because of the existence of the models for non-Euclidean geometries, in which that axiom is not satisfied. For further study of the independence of certain axioms of set theory, the student is again referred to Cohen and to Rosser (2). Also the student's attention is drawn to the series 'Studies in Logic and the Foundations of Mathematics' (North-Holland) where several books on topics of this kind are to be found.

I.7. THE INCOMPLETENESS THEOREMS AND SEMANTICS

Gödel's theorems also have consequences for semantics. For, suppose that we would formulate (analogously to what was done in Section I.6., for Syn(Z)) the semantics of Z, denoted Sem(Z); i.e., we assume the existence of one or more conditions which, within Z, express the concept 'Gödel number of a true statement in Z'. We can formulate, analogously to the proof of Gödel's first theorem, a semantic incompleteness theorem.

Before doing this, we first formulate Gödel's proof in another form. Define in Z a formula $S(x)$ containing one free variable x with the following property: (this can be done because Syn(Z) can be interpreted in Z)

(#) *$S(n')$ is true if and only if n' in Z is the name of a Gödel number n of a formula $U(x)$, which only contains x as a free variable, and for which $U(n')$ is not provable in Z.*

Let s be the Gödel number of $S(x)$. Assume that $S(s')$ is provable in Z; then $S(s')$ is certainly true, and then s must be the Gödel number of a formula $U(x)$ with the two properties as in (#). Also, the formula must be $S(x)$ itself, for each number is the Gödel number of only one formula; hence $U(s') = S(s')$ is still not provable, though, according to the above property (#), $S(s')$ is true. (Note the analogy with Cantor's diagonal method.)

In a complete analogous fashion, we obtain for Sem(Z) that, within Z, a formula $S(x)$, with one free variable, can be found with the property (cf. also Beth (2), p. 335ff.):

(##) *$S(n')$ is true if and only if n' in Z is the name of a Gödel number n of a formula $U(x)$, which only contains x as a free variable, and for which $U(n')$ is not true in Z.*

Of course, for the definition of this $S(x)$ we must use the fact that Syn(Z) is already interpreted in Z. Let s be the Gödel number of $S(x)$. Further assume that $S(s')$ were true in Z. Then s must satisfy the property (##), with the result that $S(s')$ is again not true; a contradiction! Assume that $S(s')$ is not true; then $S(x)$ must have the property (##), and then s must be substituted for x in $S(x)$, making the formula $S(s')$ true. Again a contradiction.

Our general conclusion is this: the concept 'Gödel number of a true statement' cannot be defined in Z! Besides, if we denote by Met(Z) the two axiomatized theories which we have denoted by Syn(Z) and Sem(Z), then the material of this section, along with the previous one, justifies the conclusion that *Met(Z) cannot be encompassed in our formal theory Z, both in the force of its proof as in the power of expression.*

Thus, in other words, if we assume that a system (at least as powerful as Z) be consistent, then Sem(Z) cannot be interpreted in Z. Or, in Tarski's terms: if a formal language is 'semantically closed', then it is contradictory. If we wish to introduce an acceptable definition of truth for a formal language, then we must formulate it in another language, the so-called meta-language. The meta-language is essentially richer than the object-language, i.e., it must contain variables of a higher 'logical type' than the variables of the object-language. In this meta-language we can speak about the first language – the object-language – for which we want to give a definition of truth. When the object-language and the meta-language are identical, such as is the case with natural language, then we have a semantically closed language. In such languages paradoxes appear; e.g., the paradox of the liar: "Epimenides, the Cretan, says that Cretans always lie" (see, however, Quine, p. 133), and in similar fashion the paradox of Grelling. Another paradox of the same character, but dealing with the theory of definition, is that of Berry: "the Berry number is the first number in the natural order of the natural numbers which cannot be defined by means of a sentence that contains no more than fifty words in the English language". However, it can easily be verified that the sentence defining the Berry number contains less than fifty words. This paradox can be repeated in every formal system which, for instance, allows the real numbers to be written in decimal notation.

What is characteristic of all these paradoxes is that they are in the form of *self-referential sentences*. This was also the case with the sentences we constructed in logical language in order to prove the theorem of Gödel. Likewise, the sub-title of Tarski's article 'Truth and Proof' (Tarski (4)) is "the antinomy of the liar, a basic obstacle to an adequate definition of truth in natural languages, reappears in formalized languages as a constructive argument showing not all true sentences can be proved". These so-called semantical paradoxes appear because, for one thing, one identifies the concept of truth of an object-language with that of a meta-language; i.e., one makes insufficient distinction between an object-language and a meta-language.

Following Ramsey's *The Foundations of Mathematics* (1926), one now usually distinguishes between two kinds of paradoxes, namely, logical ones and the semantical ones mentioned above. The former, such as Russell's paradox, can be eliminated by suitably changing the axioms or the syntax of language concerned. The paradox of Russell (does the set of all sets which do not contain themselves as elements contain itself or not?) disclosed a contradiction in the logical system that Frege had developed. Russell sought to forestall the appearance of such paradoxes by introducing his theory of types (cf. Section II.4.1.2). However, whether a theory is free of paradoxes can never be proven on account of Gödel's theorem. One can only see to it that certain (known) paradoxes are, in principle, excluded from a theory by adequately formalizing it.

Let us return briefly to Tarski's investigations in the area of semantics. The definition of truth must be formulated in the meta-language. But which conditions must be fulfilled? Tarski postulates that the definition must be 'materially adequate' (Tarski (2), (3), p. 65). That is to say, the definition of truth must, as much as possible, correspond with our intuitive notion of truth as it is used in the natural language. This notion, according to Tarski, agrees best with the Aristotelian notion of truth, which was formulated in the Middle Ages as "adequatio intellectus et rei". Today such a theory is referred to as the *correspondence theory* of truth.

Tarski's thought, with its sharp distinction between a formal language and that which it denotes, is often called realistic. He himself claims that "the search of truth is . . . the essence of scientific activities" (Tarski (3), p. 70).

Scientific languages are constructions that stand between reality and the natural language, i.e., a scientific language has a certain separation, not only from reality, but also from the natural language. For a scientific language is distinct from the latter by its lack of universality (in the sense of all-inclusiveness). It can claim only a great measure of exactness and specific generality in some small domain or aspect of reality. Leibniz sought a 'characteristica universalis', i.e., a language that would both be universal and exact, a language which could serve us in every science. The possibility of ever realising this ideal must be denied. For a language which encompasses everything would have to be 'semantically closed' and hence engender antinomies. And the latter is what science cannot tolerate. Every scientific language 'means' only a small part of reality. The natural sciences use the language that is geared to the natural scientific reality, a part of full reality. Mathematics 'makes' or 'constructs' in part its own reality, by building new objects or things from a given 'base material' (number, continuity), to which the mathematical language then becomes geared. What is the reality for sciences such as anthropology, or the humanities? In the natural sciences this reality is more or less independent of man. This is not the case for the humanities. For then one must deal with the reality of man in his freedom. Sartre sees the essence of man in an existential freedom. If one insists on defining the object of the humanities, then one can only, according to Sartre, proceed by saying what man is not. For only of that which is constant can one speak. Via such a 'neantization-process', man remains as 'le Néant'.

Some thinkers see the 'essence' of man as reaching out above reality, and on this basis see the possibility of scientific knowledge, also with respect to man. Others speak of the personality of man as being transcendental, and believe that complete selfknowledge, and also scientific knowledge, is an impossibility for man. Yet they still defend the view that scientific knowledge of the history of man is possible. It appears to us that the semantic incompleteness theorem, described above, presents us with a means of realising the practical impossibility of a complete and consistent knowledge in anthropology. For when already the language, if it is self-referential, gives way to paradoxes, how much more so will the language which refers to man, the procreator of language, be paradoxical. We are led very naturally, as it were, to see scientific theories as 'local' thought systems which necessarily approach, in a theoretical way, a more or less well-described part of experience, by posing limits and by systematising. Alongside this, and as opposed to this, the colloquial language maintains itself as a language, direct and paradoxical, yet useful, for it serves as a vehicle for the transportation of information, to accompany the practical, active, molding of history.

I.8. REMARKS ON NON-STANDARD MATHEMATICS

In the previous paragraphs we have been dealing with what is often called 'Model Theory'. The models that have been exposed were constructed with a view to clearing up questions in Formal Logic (cf. Sections I.4 and I.5). In the process, we needed a formalisation of the elementary arithmetic (Section I.5., Definition 6), viz. the one of the standard non-negative integers. Thus, the formalisation of arithmetic served the study of Formal Logic, and the models are sometimes called

Logical Models; they do not serve to extend mathematics proper.

As opposed to the Logical Models we also have Mathematical Models. They grew out of Logical Model Theory and they extend, complete, and sometimes replace, the standard model employed in those mathematical subjects that are in line with, and extend, the classical mathematical disciplines. For the right understanding of these so-called 'non-standard models' one has to keep in mind that there are two types of formalisation in logics and mathematics. One can, in the first place, try to formalise and axiomatise a certain discipline in such a manner that the system one gets characterises one and only one 'model' that one has in mind, and in view of which the axiomatisation is enterprised. In that case one strives to give a complete set of axioms and a language that together, and hopefully in a provable way, admits one and only one mathematical model. Such an enterprise is sometimes possible; however, it is not the most frequently occurring sort of axiomatisation and formalisation. The mathematician who is interested to think generally is rather satisfied if the system of axioms that he has concocted encompasses or covers as great a number as possible of 'models' which are useful in different mathematical fields. More often than not, these mathematicians do not formalise their system completely, nor do they write down explicitly which part of Formal Logic they use. Non-standard models in mathematics then, are models, carefully construed using the elementary predicate calculus language and Set Theory, satisfying all (or a subset of the set of) axioms that govern certain standard models of Classical Analysis, Algebra and Numer Theory, notably the real number **R** and their subring **Z** of all integers. In this way new branches of mathematics are recreated, foremostly in that the classical, Archimedeanly ordered, field of real numbers **R** is replaced by a new type of field (of so-called 'non-standard' real numbers) satisfying the same axioms of **R**, with the exception of the axiom of Archimedes; much in the same way as one constructs models for non-Euclidean geometries from or inside Euclidean space. It then turns out that the new field, though not Archimedeanly ordered, retains enough properties so that, with it as a base field, large parts of classical analysis can be rewritten. We give a sketch of how this is being done.

The construction of a non-standard model for **R** goes as follows. It uses the concept of a *filter* and of an *ultra-filter*. Let I be any set; a subset D of the power set $P(I)$ of I (i.e., the set of all subsets of I) is called a filter if the following conditions are satisfied:

$$1.\ \phi \in D \qquad 2.\ A, B \in D \text{ implies } A \cap B \in D$$
$$3.\ A \in D \quad \text{and} \quad A \subset B \subset I \quad \text{imply } B \in D.$$

D is called an ultra-filter if it is maximal in the set of filters on I; i.e., if there exists no filter D' with $D \subset D', D \neq D'$. Let for every $i \in I$ a set A_i be given, and let D be a filter on I. Look at the direct (Cartesian) set product $\pi A_i\ (i \in I)$. It consists of all vectors $(f(i))$ with components $f(i)$ in A_i. Call two such vectors f, g D-equivalent if $\{i \mid f(i) = g(i)\} \in D$. Then reduce πA_i modulo this relation to get $A(D) = \pi A_i/D$. The latter is called an *ultraproduct* if D is an ultra-filter which, in addition is *free* (or *non-principal*) in the sense that the intersection of all its members is empty. The core of the non-standard theory is the following. Suppose on each A_i we have a mathematical theory T_i, formulated in terms of predicate calculus (in the literature often called *first order predicate calculus*, in order to denote that we do not admit variables, occurring in predicates and functions, to range over predicates). Then one has at one's disposal a method of carrying the product-theory on πA_i over on $\pi A_i/D$. We say that $A(D)$ is *elementarily equivalent* to $A(D')$ (D' a filter on I) if every formula provable in $A(D)$ is provable in $A(D')$ and *vice versa*.

As a particular example take $A_i = A$ for all $i \in I$, and D a non-principal ultra-filter. Then $A(D) = A^I/D$ is an *ultra power*. A property P of A is called of *first order* if it is also a property of $A(D)$. Specializing further we shall now construct a non-standard model of the real numbers **R** and show that the property 'Archimedean order' on **R** does not carry over to any ultra product of **R**. Thus, take $A = \mathbf{R}$, $I = \mathbf{N}$, the natural numbers, and D any ultrafilter on **N**, containing all sets in **N**, having a finite complement. As such D is a non-principal ultra-filter. $\mathbf{R}(D)$ is now an ultra product. Two sequences of real numbers (a_n) and (b_n) in $\mathbf{R}^\mathbf{N}$ are equivalent if and only if $\{n \mid a_n = b_n\} \in D$. There is a map $\varphi : \mathbf{R} \to \mathbf{R}(D)$ given by $\varphi(r) = (r, r, r, \ldots)$. This map is injective (trivial) but not subjective because, for example, the class of the element $d = (0, 1, 2, 3, \ldots)$ is not an image under φ (indeed, as $\{i \mid r = i\} \notin D$ for any $r \in \mathbf{R}$, we have that $\varphi(r) \neq$ class of d). Denoting the class of an element a of $\mathbf{R}^N$ by $\tilde{a}$, observe that for every $r \in \mathbf{R}$ we have $\{i \mid r < i\} \in D$, which is nothing else than saying that $\varphi(r) < \tilde{d}$, and this means that $\mathbf{R}(D)$ is not Archimedeanly ordered. One has, on the other hand, $-\tilde{d} < \varphi(r)$, for all $r \in \mathbf{R}$, as one easily sees (note that $-d = (0, -1, -1, \ldots)$). We can now prove that the property '*Archimedeanly ordered*' (*denoted by* Arch) *is not a first-order property for the structure of fields.*

[We have to explain this statement somewhat further before sketching the proof. First of all, the structure $\mathbf{R}(D)$ is a *field*. To check this one needs all the defining properties of D to be a non-principal ultra-filter. For instance, addition and multiplication of elements in $\mathbf{R}(D)$ is defined by taking the componentwise addition and multiplication in $\mathbf{R}^{\mathbf{N}}$, and then one checks that the relation mod D respects these operations, and the axioms for a commutative ring can be verified. Then it remains to show that a series $(\tilde{a}_n)$ is invertible if it is not equivalent to the 0-series $(0, 0, \ldots) = (\tilde{0})$. The inequivalence of $(\tilde{a}_n)$ and $(\tilde{0})$ is the same thing as saying $\{n \mid a_n = 0\} \notin D$. Using the maximality of D it is now readily shown that the series $(\tilde{b}_n)$ with $b_n = a_n^{-1}$ if $a_n \neq 0$, and $b_n = 0$ if $a_n = 0$ is the inverse of $(\tilde{a}_n)$ in $\mathbf{R}(D)$. Secondly, to show that Arch is not a first order property, one needs one more lemma that will not be shown here. It says that if A is an ultra power then for any set I the power A^I is also an ultra power which is elementarily equivalent to A.]

Thus, assume that in predicate calculus there exists a sentence ARCH expressing that the property Arch is a first order property. Then in any model A, that is Archimedeanly ordered, ARCH is provable and *vice versa.* In particular, taking $A = \mathbf{R}$, we get that ARCH is provable in $\mathbf{R}$. However, $\mathbf{R}(D)$ is an ultra power of $\mathbf{R}$; hence 'ARCH is provable in $\mathbf{R}$' is the same thing as saying 'ARCH is provable in $\mathbf{R}(D)$', which is a contradiction, because $\mathbf{R}(D)$ is not Archimedeanly ordered, as we have seen. In the same manner one can prove that there cannot be given finitely many axioms which describe fields of characteristic zero. In the same vein, the property 'well-orderable' from set theory turns out not to be a first-order property. For further literature on non-standard mathematics the reader is referred to Luxemburg, Robinson and Hurd and Loeb. From the above construction the reader can gather that non-standard mathematics is not a constructive affair, i.e., even the non-standard integral numbers $\mathbf{Z}^{\mathbf{N}}/D$, D as before, cannot be constructed by any procedure which resembles the step-by-step construction of the integers; just because there is no recipe known which enables one to, in a step-by-step fashion, collect into a set a non-principal ultra filter D on $\mathbf{N}$ (cf. also Section II.3).

Chapter II

Epistemological Aspects of Mathematics in Historical Perspective

-ξὺν νόωι λέγοντας ἰσχυρίζεσθαι χρὴ τῶι ξυνῶι πάντων, ὅκωσπερ νόμωι πόλις, καὶ πολὺ ἰσχυροτέρως - Heraclitos, DK B frgmnt 114.

[Those who speak with reasonable insight should reinforce themselves with that which is common to all, just as a town becomes strong by its legislation, and even stronger (than a town).]

II.1. INTRODUCTION

Having gone into some detail on the structural (i.e. semantical and syntactical) moments of a mathematical theory, we shall now broaden our scope and show how mathematical theories are related to more general philosophical, especially epistemological but also some ontological, questions. This broadening of the terrain will require, however, some knowledge of questions of a philosophical nature. To plunge into the present discussions on the philosophy of mathematics immediately would be somewhat premature. Our procedure shall be first to look back in time and see how the present problems have arisen, thereby allowing ourselves to gain valuable insight into the origin of present discussions, while at the same time letting related philosophical problems come to the fore. In this manner the discussions on the philosophy of mathematics will be seen against their historical background while the related philosophical questions will lose some of their unfamiliarity.

II.2. THE PHILOSOPHY OF MATHEMATICS IN HISTORY

II.2.1. Greek Mathematics

The philosophy of mathematics has no distinct history of its own. Ever since the early Greek thinker and mystic Pythagoras (±570 - ±500 B.C.) injected number and figure as necessary elements into the methods of speculating about the structure of the world, the fundamental entities of mathematics and the mathematical method have made for themselves an irretractable position in philosophy. Not every philosopher, however, sees this influence as a happy one. Whereas philosophers such as Plato, Leibniz, Kant, etc. are open to mathematics, men such as Aristotle, thc scholastics, Hegel, Heidegger, etc. denounce its decisive influence on philosophy as having led the latter onto a wrong track. Before mentioning further names and historical movements, we do well to look first at the type of philosophical problematics that mathematics was called upon to aid, which problems, conversely, exerted their influence on that which men thought mathematics to be all about.

The most important problem of philosophy that mathematics was supposed to aid in, was the problem of constancy and change. Not only is this problem important but also it is among philosophy's oldest problems. Both poles of this problem had their defenders among the earliest Greek thinkers. *Thales* (±624 - ±545 B.C.), to name a proponent of one side, concluded that everything *is* water, stressing thereby not that water was a fluid and, hence, could flow and change, but that the make-up of everything was to be seen from out of the element 'water'. There was a structure in all things and that structure was best accounted for, according to Thales, when taken to be (modifications of) water.

The champion of change and flux was *Heraclitus* (±540 - ±480 B.C.). 'All things flow' was his adageum. The world for him was in ceaseless flux; nothing *is*, only change is real. *Being* had for him a dynamic character which, for Thales and, after him, Parmenides and Zeno (the one who formulated the paradoxes, attempting thereby to prove that change had no positive being!) it could not have. The pressing epistemological side of this dilemma is: Where and how is true knowledge to be found and obtained if the knowable is ephemeral? The heart of any ontology (= science of being) is here laid bare, together with its consequences for epistemology (= science of knowledge).

The place of mathematics in this problematics will become clear if we first take note of a development that took place after the rise of the stated dilemma. The earliest Greek thinkers had treated the world and its beings in itself. The beings were composed of the four (later five) universal elements – earth, water, air and fire (later ether was added) –, and all the modifications of beings had to be accounted for in terms of these elements. But a new movement came up. This new movement did not so much reject the thoughts of its predecessors and contemporaries as that it introduced a *new* 'entity' into their discussions on being. What was added were the *qualities* – the warm and the cold, the dry and the wet, etc. –. A being, for the thinkers of this movement, was not simply an aggregate of cosmic elements in some proportion, but also had qualities running through it which served to enhance its being and its knowability. Qualities endure whereas beings pass away. It was within the problematics of constancy and change that the qualities became much more prominent than the cosmic elements. Qualities became the structural components of things.

Within this thought-climate the philosophy of mathematics took its roots. Among the ranks of the thinkers stressing quality as opposed to 'naked' being, a dispute arose with regard to the following question: Is it sufficient for the knowledge and determination of things if they are seen only in conjunction with the infinitely varying, unspecified qualities, or must some boundary and measure be applied to the infinite variety? This problem divided the movement. Some felt that one could only account for things sufficiently when seen in their *unspecified quality* (e.g. Parmenides, Anaximenes), while others (Pythagoras and his followers) thought it necessary to place the *infinite qualities within bounds.* It was to accomplish the latter that numbers and geometric forms were introduced, i.e., as *metric* determinations of indeterminate qualities.

Now, as never before, a speculative role was fulfilled by mathematical entities. Qualities can be of degrees, compared as to intensity, and hence can be determined by number. Furthermore, colours have the added property that they also define the boundary or surface of things. The quality 'sound', in all its degrees, was found to be harmonic or unharmonic depending on the length of the string and its thickness, while under constant tension. Here a very simple and awe-inspiring application of number in music was found. In short, the qualities the philosophers had looked for, in order to come to an abiding knowledge of the nature of things, were many times increased in number by this 'quantification of the predicate'. And it is for this reason that speculative thought took to include and interpret the mathematical entities, qua geometry, into its philosophical domain. This mathematically determined philosophy was called *Pythagoreanism* because Pythagoras, and his school, brought it into philosophy.

Note that in Greek mathematics, the bridge between mathematics and philosophy was a geometrical one. This may be explained by the fact that number was seen to be subservient to form. Number aided form in denotation of length and in calculations of length, areas and volumes. Numerical relations themselves were investigated via geometrical forms. In the words of Heath, an outstanding authority on Greek mathematics: "With rare exception, . . . the theory of numbers was only treated in connection with geometry, and for that reason only the geometrical form of proof was used, whether the figures took the form of dots marking out squares, triangles, gnomons, etc., or straight lines" (Vol. I, p. 16). Furthermore, since philosophy demanded metrical determinations of qualities, which, in turn, determined extended bodies, the mathematics of that day was able to fulfill this demand very adequately. Another reason for this 'geometrising of arithmetic' lies in the discovery of incommensurate magnitudes. We shall touch upon this later.

The 'birthday' of the philosophy of mathematics, brought about by Pythagoreanism, must be looked at somewhat more closely in order to surmise its meaning and consequences. No one will minimize the fruitfulness of the newly formed alliance, as history has abundantly shown. The bond between philosophy and mathematics enabled each to exert a stimulus and an influence on the other. For instance, the Academy of Plato became a center for mathematicians and philosophers alike. But with all due recognition of this fruitful cooperation, there is also a serious negative side to it all. Whereas pre-Pythagorean mathematics was based entirely on practical reckoning – counting and measuring –, and found its meaning there, the Pythagoreans transformed this meaning and method to something quite different. Mathematics was taken out of the world of practice – of commerce, for instance – and made to see itself subservient to speculative thinking. Because the mathematical entities were interpreted as final and ultimate determinations of the world, a conscious *separation* was made between pure and applied mathematics. The latter continued to be 'in' the world, but the former, pure mathematics, because its entities had such an ultimate ontological status, could not remain 'in' the world. When discussing Plato we shall see how this has been accentuated and developed further. This separation of pure and applied mathematics contributed to the disturbance of the intrinsic development of mathematics as an interplay between Geometry and Number Theory.

We must not forget the type of thinking mathematics was encased in. There resulted, first of all, as we already mentioned, the preference for geometry with its forms to that of arithmetic (and algebra) with its numerical operations. We may certainly say that *the Greeks have given the Western world geometry, but algebra and the number system has had to come from the Indian and Arabic world.* There is every reason to uphold that this rejection of algebra by the Greeks was consciously done, as not 'fitting' the philosophy they subjected mathematics to.

Secondly, there is the necessary *metrical* character of the geometrical forms, and consequently, the recognition of only Euclidean metrical geometry, that held up the development of mathematics as a pure science. It needs no mentioning how long it took before this condition was first made relative – in the mid-19th century with the rise of non-Euclidean geometry – and later its priority abandoned: in projective geometry and topology. The consequence of this, for the view of space that the modern world has inherited, is this: with the waning grip of metrical geometry on space, we can see more clearly the difference between, on the one hand, geometry and its constructive character, and, on the other, the continuum and its non-constructive character. We return to this question later (III.1.).

But let us follow the track of the Greeks and their philosophy of mathematics. The next important development after Pythagoreanism in the philosophy of mathematics is that which Plato (428/7 - 348/7 B.C.) introduced. We can do no better than to quote one paragraph from Aristotle's *Metaphysics* as to what this introduction was, how it related to what had gone on previous to Plato, and what an inherent difficulty lay in this position.

For, having in his (= Plato's) youth first become familiar with Cratylos and with the Heraclitean doctrines (that all sensible things are ever in a state of flux and there is no knowledge about them), these views he held even in later years. Socrates, however, was busying himself about ethical matters and neglecting the world of nature as a whole but seeking the universal in these ethical matters, and fixed thought for the first time on definitions; Plato accepted his teaching, but held that the problem applied not to sensible things but to entities of another kind – for this reason, that the common definition could not be a definition of any sensible thing, as they were always changing. Things of this other sort, then, he called IDEAS, and sensible things, he said, were all named after these, and in virtue of a relation to these; for the many existed by participation in the Ideas that have the same name as they. Only the name 'participation' was new; for the Pythagoreans say that things exist by 'imitation' of numbers, and Plato says they exist by participation, changing the name. But what the participation or imitation of the Forms could be they left an open question. (Book A, Chapter 1, 987 à 31 - b 13)

It is clear from this quotation that Plato was concerned with the foundation of true knowledge, which foundation he –

and all the pre-Socratic thinkers – sought in the nature of things (ontology). But here Plato introduces his new element: those entities about which we have true knowledge cannot be 'in' or 'on' the things of the sensible world, because the latter are in constant flux. Nor is it sufficient to seek this constancy in the metrical characteristics of the qualities, as the Pythagoreans did, since by that method we can attain only relative knowledge of relative metrical relationships. Combining with mathematics the study of man (ethics), Plato came to assume "the existence of absolute beauty and goodness and magnitude" (Phaedo, 100b), in a world accessible only to the intellect, and which was, at the same time, the foundation for the things in the sensible world. The relation of the latter to the static absolute entities in the intelligible world was one of 'participation'. The Good Itself, the True Itself, and the Beautiful Itself – as the Ideas were later called by Plato – were the basis for the existence of the human soul; the Numbers Themselves and the Geometrical Forms – as the mathematical magnitudes were later called – founded the sensible world and also man's own bodily existence. Because the intelligible entities have *real* existence for Plato, in a world distinct from the sensible world, his philosophical doctrine has received the label 'realism'.

Two more factors should be mentioned which will aid in seeing Plato's reasons for taking this position. Before becoming a realist, he had already accepted the doctrine of the absolute dualism between body and soul. The body was for him the prison of the soul, the source of evil in the world; the soul was pure, immortal, the principle of all good. Furthermore – and this is the second point – Plato had developed an epistemology which took into account both his anthropology as well as his battle with the Sophists of his day. Man's knowledge, Plato defended, is basically a *remembering.*

Combining these two factors, we see that, because the soul is immortal, it must have existed before it was incarcerated within the body. But the body being the principle of evil, ignorance resulted in the soul because of this imprisonment. The soul, having had true and certain knowledge before birth, must now, in life, regain this knowledge by effort. It must dispel the influence of the world, the body and the senses, in order to recall something of its previous state. But where was that state? This Plato did not answer in his earlier dialogues (e.g. Gorgias, Menon), but his becoming a realist allowed him to answer: the soul recalls its previous state when it is in the realm of the pure intelligibles – the Ideas, the Numbers Themselves and the Forms. For the rest of his life Plato remained a realist, though his view of the structure of man and the sensible world was to go through many phases yet. But these not being directly relevant to the purpose at hand, we let them rest.

Plato's position is very important, historically speaking, because subsequent thinkers, with a small revision of Plato's realism, introduced the doctrine that man has *a priori* knowledge. It is this doctrine that has exerted a great influence in forming the modern mind, including the discussions on the philosophy of mathematics. But more of this later.

Before passing on to Plato's most important pupil, Aristotle, we must mention a further development in Plato's view of mathematical entities. Towards the end of his life, in a lecture entitled 'On the Good', Plato tried to soften the strict separation between the sensible world, with its sensible objects and sensible determinations (this part being identical with the Pythagoreanism of his day), and the intelligible world with its Ideas, Numbers Themselves and Forms. The separation of these two 'worlds' carried the problem as to how sensible instances of, say, two-ness or circularity, could all participate in the single ideal entity 'two itself', or 'circle itself'. For the sensible things are temporal, changeable and many, while the intelligible entities are eternal, unchangeable and unique. Clearly, the two kinds of entities have entirely disjoint properties. On the testimony of Aristotle, we know that Plato introduced a third kind of entity, which he called "objects of mathematics, which occupy an intermediate position, differing from sensible things in that there are many alike" (Metaph. A, Chap. 1, 987 b 15). It is here that the subsequent Frege-Russell definition of number, as will be discussed later, finds its inspiration.

Aristotle (384 - 322 B.C.) found this entire view unsatisfactory. He wanted to break radically with the separation of an intelligible world from that of the sensible world. Somehow, what mathematics, or science in general, dealt with had to reside 'in' the sensible things, though not in the way previous thinkers before Plato had thought. Accordingly, prompted by the results of his investigations in biology, he came to see things as a substance, a unity of two components, namely matter and form. 'Matter', in itself and pure, was entirely formless and inert; 'form' had the character of an active principle and was called an *entelechy*. This 'form' concept was a teleological concept, for, in the things, the entelechy was the principle that guided matter to its appropriate end (Gr: telos) or purpose. E.g., the seed of an oak tree becomes itself an oak tree because the matter composing the seed has as its form an internal principle that will guarantee that the seed will indeed become an oak tree; in short, it has the form 'oak tree'. We see that this 'form' is not the Pythagorean-Platonic notion of *shape* but rather that of *purpose*.

On this teleological basis, it is clear that the mathematical entities will be hard to define. Since only that which is *in* things can be spoken of and thought about, mathematics must then deal with approximations of things insofar as mathematics deals with sensible things. Accordingly, this aspect of Aristotle's view of mathematics was, in later history, to be called an *a posteriori* or empiricistic view, in contradistinction to the more platonically inspired *a priori* views.

With Aristotle's minimizing of the ontological implications of mathematics, the epistemological and formal-logical side of it, received automatically much more emphasis. As founder of the science of logic, he devoted much attention to the structure of science in general, and to the structure of whole scientific theories. He clearly distinguished between:

(i) the principles which are common to all sciences (axioms or common notions);

(ii) the special principles which are taken for granted by the mathematician engaged in the demonstration of theorems (postulates);

(iii) the definitions, which do not assume that what is defined exists (definitions); and

(iv) existential hypotheses, which assume that what has been defined exists (unit (monad) and magnitude, points and lines – everything else must be constructed!)

Hence, it is in the period of Greek thought that we find the first formulations of fundamental questions regarding mathematical theories. Aristotle was not unique in this respect, for, within the Academy of Plato, there had arisen discussions and opinions on this subject among Plato's immediate followers. There were discussions as to whether mathematics concerned itself only with *theorems*, i.e., deductions (or, more literally, investigations) – as argued by Speusippos – or whether only *problems*, i.e., constructions, were proper to mathematics – as argued by Menaechmos. Others gave place to both theorems and problems, as we find slightly later in Euclid's case. Also in this period the components necessary for a complete proposition (a theorem or a problem) and its proof were explicitly stated. No less than six parts were distinguished: enunciation in general terms, setting-out of particular data, definition (or specification) of general problems in terms of the specific data, construction, proof, and conclusion. Furthermore the method of proof was spelled out. A proof was always *synthetic*, i.e., it proceeded from the known to the unknown, from the simple and particular to the more complex and general. The opposite process, i.e., proceeding from the unknown and assuming it to be true, then reducing it to simpler propositions, stopping only when a known true proposition is arrived at, was called *analysis*. Analysis proper had no place in the proof of a theorem, though it could play an important part in the discovery of the proofs. Analysis gave, at most, an indication that the proposition was true, but it was not acceptable until a proper synthetic proof was given. (For more details, see Heath, Vol. I, p. 370, 66)

This suppression of analysis as acceptable mathematics has very interesting consequences. With the exception of a manuscript of Archimedes (found only in 1906) there is no clear account of how the Greek mathematicians actually carried out their analysis while searching for proofs of their theorems. The analysis given in some of Euclid's proofs play only the minor role of determining some of the data involved in the theorem. But every analysis had to be back-tracked by the method of synthesis. Archimedes' discussion of the method of analysis he used, is found in a letter addressed to Eratosthenes. Archimedes also states that this method only indicated that a theorem may be true, the actual proof must follow synthetically. What is unique about this method is that it employs mechanics for its analysis. The unknown is made to 'balance' with a known factor, thereby 'reducing the unknown to the known. The trick was, to use infinitesimal elements in making the reduction, so that, in the synthesis that followed, the method of exhaustion – the precursor of the integral calculus – could be used. Because the primitive 'differentiation' here used was a method of analysis, no further development was made of it until Newton and Leibniz took it up again.

Incidentally, for the Greek mathematician, the well-known *reductio ad absurdum* method was a variety of analysis, so that it too, could not function as a positive synthetic proof (Heath, Vol. I, p. 572).

This predominance of the synthetic method had to do with the Greek fear of the infinite. This scare came on them when the Pythagoreans discovered the irrational magnitude $\sqrt{2}$. Their work with numbers had convinced them that between any two whole *numbers* there was always a common unit, hence, they were commensurate. When it was found that the relation between *magnitudes* was not always commensurate, i.e., no matter how far one subdivides, never will a common magnitude be found, working with spatial lines via their lengths (numbers) had to be given up. The only way to handle magnitudes was to treat them as segments and from there to make further constructions with them. Any pending 'arithmetisation' of geometry was radically cut off. This had a further repercussion in their method of proof. We know that early theorems were proven by analysis, but with the discovery of incommensurate magnitudes, this method became inadequate. For, it was now possible that analysis would not be able to reduce the unknown, assumed to be true, to something known, because the analysis may have to be carried on *ad infinitum*. The synthetic method, overcoming the unknown from out of the known, became the only reliable method in geometry. When notions such as "the infinitely great and the infinitely small" came up, they were usually substituted by: "greater or less than any assigned magnitude" (Archimedes, p. cxlii).

With this said, we take a last look at Aristotle and see what light he brought into this question of the infinite. In the third book of his Physics, he makes a distinction between the infinite as a process, i.e., as an increasing without end or a subdividing without end, and the infinite as completed totality, i.e., as actual infinity. We have here a clear distinction between the *potential* and *actual infinity*. Since precisely this point has become perhaps the main bone of contention among 20th century mathematicians, we quote his argument for retaining the former while rejecting the latter.

> But my argument does not anyhow rob mathematicians of their study, although it denies the existence of the infinite in the sense of actual existence as something increased to such an extent that it cannot be gone through; for, as it is, they do not even need the infinite or use it, but only require that the finite (straight line) shall be as long as they please . . . Hence it will make no difference to them for the purpose of demonstration (Physics III, Chapter 7, 207 b 27).

We have dwelt on the Greek period of the history of the philosophy of mathematics at some length, since many of the present problems in the philosophy of mathematics have their origin here or were touched upon by the Greeks. Questions concerning the status of infinity, the ultimate adequacy of analysis for mathematics, the place of construction and synthesis, the nature of mathematical theories, the meaning of mathematics for philosophy and for the certainty of science and knowledge, etc. are all themes which are to recur in later centuries.

II.2.2. From Hellenistic Philosophy to Modern Rationalism

Whereas Hellenic (classical Greek) philosophy was exceptionally fruitful for mathematics and the reflection upon mathematics, the succeeding centuries witness a much slower development. Part of this period, the so-called Middle Ages, shows next to no interest, let alone development, in mathematics at all. For the continuity of our story, nothing is lost if, after mentioning what happened in the Greek philosophical schools of the Hellenistic (i.e., post-Hellenic) period – in so far as it is relevant to our subject, of course – we jump immediately to the modern period of Western thought.

The most important development in post-Hellenic philosophy, and consequently for the philosophy of mathematics, is the rise of the theme of *a priori* knowledge. With the passing of the classical Greek period, a wave of scepticism and even agnosticism swept over the philosophical schools. They began first to doubt the knowability of the entities Plato had posited in his intelligible world; then, stronger yet, they denied their existence. This had drastic practical consequences: the Ideas Plato posited as abiding entities (the Good etc.) necessary to govern human conduct, as well as the metrical limits (Numbers Themselves and geometrical forms) as the foundation for objects which science (mathematics) ultimately referred to, were now discarded. Society had no values and science no basis! In overcoming this crisis, the notion developed that the Ideas and mathematical entities must be of a conceptual nature and hence must reside, not *outside* of the human mind – as the classical Greek thinkers had always maintained – but *within* the mind, as an integral part of its inborn structure. The fundamental knowledge of the Ideas etc. thus came to be seen not as gleaned from experience – so that it could be doubted, or even denied – but in fact, as *constituting* indubitable experience itself.

It is obvious that such a transformation did not occur overnight. To reconstruct it would not contribute to our topic, hence, for the most part, we pass it by. What is important in this is the fact that via this theme of the *a priori*, modern thought, from Galileo to the present, has been most decisively influenced. Before proceeding to this period, we will mention briefly how the Pythagorean school fared in the transformation referred to.

The Pythagorean, now to be called the *neo*-Pythagorean, school also followed suit in making *a priori* numbers and metrical forms. To come to explicit, conscious knowledge of these entities, one needed not instruction (= tuition) from without, from the outer world. One had merely to turn inwards to be instructed in ultimate mathematical knowledge. In a very literal sense of the word, one came to mathematical knowledge by an in-tuition, by an act of self-reflection, of self-instruction.

This made it possible to combine a religious mysticism with positive, scientific work. Whereas for the early Pythagoreans, numbers and forms constituted a world, these entities now attained a more spiritual function. Hence number-mysticism and number-fantasy found a ready home here. The school had such an appeal that for centuries it was able to exert an influence. We find the Renaissance filled with elements of it – the cabalists, for instance – and Kepler himself loved to indulge in it. Nevertheless, the modern (rationalistic) thinkers saw it as their task to rid mathematics of its mystical elements and to retain its scientific content. This they did by combining mathematics with physics, thereby bringing mathematical astronomy to the foreground again, as opposed to astrology, as well as by developing a theory of mathematical motion. By rejecting the cabalistic and alchemistic tendencies that had also sprouted up under its tutelage, neo-Pythagoreanism was undone of its mystical and spiritual force.

II.2.3. The Period in Early Modern Philosophy

What makes modern philosophy the force that it is, is its rationalism. Rationalism is the doctrine that defends the notion of *reason* and reason is human understanding along with, and inherent in it, *a priori* knowledge. In all cases the *a priori* knowledge includes, in some measure, the mathematical notions of number and spatial figures. The spirit of this kind of philosophizing and the relation of philosophy to mathematics it implies, is well caught in the words of E. W. Beth, who approvingly said: "De beantwoording van het probleem van de grondslagen der wiskunde vooronderstelt of sluit in de beantwoording van de hoofdvragen der wijsbegeerte. Wie het wezen van de wiskundige denkwijze heeft begrepen, begrijpt het wezen van den denkenden geest" ((3) pp. 91-92; translation: The answer to the problems in the foundations of mathematics presupposes and includes the answer of the principal questions of philosophy. Only those who have understood the essence of the mathematical thinking pattern, understand the essence of the thinking mind).

We shall now discuss how this and other salient points in the philosophy of mathematics were formulated by some individual thinkers.

II.2.3.1. Descartes (1596-1650)

Descartes was one of the first to project a philosophy and cosmology that was patterned according to the mathematical way of thinking. The essence of things he saw in their extension, and the modifications of them – figure, motion, momentum – as functions of extension. The qualities of things, colour, temperature, hardness, etc., which, from days of old, were always considered to be inherent in things, were now, via Galileo who first brought about this change, considered to be figments of the human subjectivity. Accordingly, the doctrine of *primary* and *secondary qualities* was stated, in which the mathematical properties of things were considered primary and essential to the structure of things, while the sense qualities were considered secondary and nominal. Descartes saw all things – except the human soul and God – as characterized by the mathematical property 'extension'. Because extension was taken to be primary and fundamental, dependent on no other properties, Descartes saw it as a substance. His physics and his cosmology were developed entirely on the basis of geometry.

God and the human soul (mind) formed another substance, distinct from the substance extension. Their essence was to *think*, whereby Descartes understood all doubting, understanding, affirming, denying, willing, imagining, perceiving, loving and hating (See his 2nd and 3rd Meditations).

Thinking that most of these faculties of the mind lead to confused knowledge, and wishing to come to a circumscription of clear and distinct knowledge, Descartes brings a bifurcation into the mind. This is done by the *intuition*. Of this he says:

By intuition I understand, not the fleeting testimony of the senses, nor the deceptive judgement of the imagination with its false constructions; but a conception of a pure and attentive mind so easy and so distinct, that no doubt at all remains about that which we are understanding. Or, what comes to the same thing, intuition is the undoubting conception of a pure and attentive mind, which arises from the light of reason alone, and is more certain even than deduction. (Rules for the direction of the understanding; Rule III)

It is this intuition, as the undoubting conception of a pure and attentive mind, arising only from the light of reason, that Descartes wished to explicate and distinguish from the other sources of (not-so-clear) knowledge. The resulting theory was to be a *mathesis universalis* (mathesis here means: theoretical knowledge in general). Its most important components were to be arithmetic, geometry, music, astronomy and physics. The Discourse on Method was the closest realization of this plan.

In this work Descartes gave a long defence of the existence of the self (mind) and the sure and indubitable knowledge we have of it, even when we are under compelling pressure to doubt it. The methodical doubt was used by Descartes to come to a knowledge of the self that could not be doubted. Doubt vanished only when it had arrived at "the undoubting conception of a pure and attentive mind". From this *one* starting point, this sure 'Archimedean point' (see 2nd Meditation), where only the sure light of reason shone, human knowledge was to depart. Appended to the *Discourse*, as exemplifications of this approach, Descartes gave the public his work on (analytic) geometry, optics and meteorology.

The intuition as the starting point for thought is supplemented by Descartes with another mode of attaining knowledge: *deduction*. Deduction is not on a par with intuition, because deduction is always in need of the intuition and not *vice versa*. Deduction is that procedure whereby ". . . many things are known with certainty, which nevertheless are not themselves evident, simply because they are deduced from true and known principles by the continuous and uninterrupted movement of a mind which clearly intuits each step". (Rules, Rule III)

This entire method of Descartes was clearly patterned after the mathematical sciences: arithmetic and geometry. It is here that he found concepts clear and distinct, because the fundamental concepts of these sciences, or 'the first seeds of truth', were 'sown by nature in the human mind'. We have here the clear statement of the *a priori* tradition Descartes wishes to continue in.

The discovery of analytic geometry hangs very much together with Descartes' methodical monism. By starting from one intuitively clear point and expanding it so as to include all the sciences, Descartes was able to emphasize the interrelatedness of algebra and geometry. By taking the best traits of both these subjects together, not only were the faults of the one corrected by the other, but also the science of *analytic geometry* was born.

It is clear that with respect to the alternative mathematical methods, namely, analysis or synthesis, Descartes chose for the former. The synthetic or axiomatic method, being a deductive method, was but a *help* to build and expand intuited knowledge. The result is that no theorems are ever mentioned or proven in Descartes' geometry. He works entirely on problems and constructions which, when assumed to be true and analysed into their clear and distinct parts and relationships, needed no elaborate synthetic method to confirm their truth, as in Greek geometry. Descartes' intuitionism and Platonically oriented rationalism made him see mathematics as an *a priori* science and to prefer the analytical to the axiomatic method.

II.2.3.2 Newton (1642-1727)

Isaac Newton will briefly be mentioned so that we can better understand Leibniz, who stood antithetically to Newton, and Kant, who tried to overcome this antithesis by a transcendental method of reasoning.

Though Newton cannot be considered to have a strict philosophy of mathematics, his views of space and time are of considerable importance. Not only does he offer one of the alternative views with regard to these concepts as the foundation for mechanics (as opposed to Leibniz, about whom later), but also the character of analysis, being developed in correlation with physical questions, is involved. Since mathematical analysis, which deals with 'the infinite', and which, via the marvelous development of the integral and differential calculus, found direct application in physics, the *status of the infinite* was directly concerned. Certainly, the Greek 'fright' for the infinite was gone, but conceptual clarity remained lacking. Newton, as we shall see, brings his notions of space and time in conjunction with the being of God; Leibniz turns to symbolic logic for grasping, be it symbolically, what only God can actually come to know. But we are running ahead of ourselves.

Newton's work shows a distinct difference from that of Descartes. Whereas the latter started his science from *a priori* concepts which one comes to know through an intuitive self-reflection, Newton proceeds on empirical grounds. He makes a fundamental distinction between the phenomena and the 'things themselves', and explains their relation as that of effect to cause. Phenomena must be *described*, while the things themselves would offer *explanations* for the phenomena. All science (= natural philosophy) must concern itself with the phenomena. In this domain we perceive qualities such as hardness, extension, impenetrability, moveability and inertia, and by *induction* we conclude that these qualities apply to all bodies. Newton explains that his laws of motion were found in this fashion. They were inferred from the phenomena and made general by induction. They are nothing but descriptions of what is observed and do not mean to explain anything (cf. also Bunge).

To explain things, one has to frame *hypotheses* as to the nature of the things themselves and this was clearly contrary to the method of science as proposed by Newton. 'Hypothesis non fingo' was his adage. However, Newton himself certainly tried to formulate hypotheses which would account for the phenomena. These 'speculations' he clearly distinguished from his work in natural philosophy, so that one need not see a contradiction here. The speculations Newton indulged in brought him to certain views on the nature of God and to various alchemistic theories.

Newton's view of space and time as absolutes prevents one from seeing in Newton a consistent empiricist. Absolute space and time are, on the one hand, to be distinguished from relative space and time, which can be perceived with bodies, and on the other hand, are necessary notions for physics. The absolutes are neither empirical – or else they would be relative – nor hypothetical – or else they would not belong to science. His solution was to connect them with the being of God, so that space becomes the *sensorium Dei.* What this actually comes down to is that absolute space and time are *a priori* structures, necessary to come to any empirical or hypothetical theories concerning the structure of things.

II.2.3.3. Leibniz (1646-1716)

In Leibniz's mature thinking we find a very closely knit

system in which metaphysical, logical, physical, biological and mathematical ideas all play a vital role. Something must be said about all these points if Leibniz's thought is at all to become intelligible and if his philosophy of mathematics is to find a background and relief.

Leibniz rejected atomism, though in his youth he had defended it. It is incomprehensible, he says, how things, which are distinguished only by number (i.e., two things, three things, etc.) and for the rest are entirely identical – as are the atoms, can be spoken of at all. For this is entirely contrary to reason. There must be a *sufficient reason* for making a distinction between things. To say that there are two leaves and then to consider the leaves otherwise identical is, in fact, to fail to distinguish them at all.

Two ontological principles are at work here: the *principium identitatis indiscernibilium* (the principle of the identity of indiscernibles, i.e., things are identical if no difference is discernible) and its corollary, the *principium rationis sufficientis* (the principle of sufficient reason, i.e., there must be a sufficient reason for things to be as they are in distinction from other things as they are). Hence each thing which is distinct from another thing must have an *intrinsic* difference with respect to that thing.

Leibniz generalized this into a general principle of individuality. Every thing, the non-living included, must have a distinct internal principle which makes it distinct from others. Organic things all have a *life-principle* which, in plants is called an entelechy, in animals a soul (psyche) and in man a rational soul or mind. These principles of organic beings make them to be *subjects* and have *perceptions*. What makes each being distinct is the degree of clarity in their perceptions. By perception is not meant a sensing of something from the outside, but an internal drive which shows itself by the degree of activity of the subject. Thus plants perceive very indistinctly and show the least amount of activity (conatus, striving). Animals have a more developed subjectivity in that they have a memory. Accordingly, they show much more activity. Man has, over and above this, ideas implanted in his (animal) soul making it a rational soul or mind. This makes him capable of having a knowledge of the eternal truths and of entering into (social) relationship with God. All these principles of subjectivity are furthermore eternal and imperishable.

Physical things are not of this character, although there is a way of viewing them as degenerate species of these life-principles. The activities of physical things are not known from an immutable, internal striving, but from *mutual* energetical interaction. Now the amount of kinetic energy in any interaction remains constant, as Leibniz had discovered, so that here he saw an analogy with the imperishability of life-principles. These constant kinetic energy patterns Leibniz called *forms*. They are not alive, though they show an analogy to life. Leibniz emphatically denies that a stone, for instance, has a soul or is somehow alive.

Leibniz's metaphysics now continues this same line. Each life-principle (or soul) is an intrinsic activity that calls forth its own perceptions. It has a substantial character and is closed to any outer influence. These substantial life-principles Leibniz called *monads*. Everything in the universe is made up of monads. Now to see the universe of monads in its coherence, as a cosmos, Leibniz postulated a *pre-established harmony*, which related each monad to every other monad. To accomplish this, Leibniz says that the perceptions of each monad are, in fact, a picture, image, or perspective of the total universe seen from the point of view of the monad in question. Thus the internal change *in* each monad stands immediately in harmony, or correlation, with the state of the universe. Each thing is a unique individual because its (dominating) monad reflects, from its own point of view, the universe, while at the same time – and this is the harmony of it all – the reflections of the universe in each monad are the perceptions that the monad *itself* brings forth (more or less clearly). This harmonious universe of monads is ultimately the perceptions, or fulminations of one central monad. And this is God.

Bodies, or objects, in contradiction to monads, or subjects, are seen to be nothing but *composites* of monads. Bodies have no real, individual character and are accordingly considered to be *phenomenal.* Bodies have extension because of the multiplicity of monads they contain. Monads themselves are extensionless. Each body is an actual infinity of (point-) monads. Extension, motion, position, time, etc., which are all modi of bodies, also have *no essential* character. They function as ways of *ordering* the monads and have no further status. Time is defined as "the ordering of what does not exist simultaneously", and space is "the ordering of what co-exists", or does exist simultaneously. This relative character of space and time brought Leibniz in complete opposition to Newton's absolute notions of space and time. Furthermore, Descartes, with his qualification of bodies by extension and then substantializing this, was also left far behind. Whereas Descartes saw the sense qualities as secondary, in a disqualifying sense of the word, and the mathematical qualities as primary, Leibniz now states that these latter qualities are also secondary and products of the mind. This was not to minimize them but to lay the emphasis where, according to Leibniz, it belongs. Also for Leibniz the focal point is the mind. But that mind is the seat of (rational) ordering, constituting experience in its entirety, hence also in its mathematical aspects.

For this reason, less emphasis is placed on the ontology of mathematical entities and more on the epistemological side of it. To this we now turn.

Faithful to his view of the mind as a monad, i.e., a subject calling forth its own perceptions (in this case: concepts and notions), Leibniz sees each logical proposition in the scheme of subject and predicate, whereby the predicate is always *contained within* the subject. Each proposition is thus an analytical proposition and its truth is completely determined according to the *principle of contradiction* or identity, i.e. (A is not not-A) or (A is A). Note well: this principle is a *logical* principle and must be distinguished from the *ontological* principle of indiscernibility, although the parallel between the two is obvious.

These 'truths of reason' (as Leibniz called the propositions that fell under the principle of contradiction) are necessary because their opposite is impossible. Mathematics is entirely accounted for in this *a priori* realm of reason. *Definitions* may, by analysis, be decomposed to their simplest ideas and, lacking further definition, they must be *a priori* given. The *theorems* and *problems* of mathematics are reduced, by analysis, to definitions and primitive axioms and postulates. These axioms and postulates have no need of proof because they are identical propositions.

Then there are also 'truths of fact'. These also are propositions in the analytic form, but their predicates would require an infinite analysis to derive them from the subject. Accordingly, these truths have no necessity in them. They are contingent and their opposites are possible. They are governed by the principle of sufficient reason. In virtue of this principle, no *fact* can be real or existing (i.e., in the ontological meaning of the principle) and no *statement* true (i.e., in the logical meaning of the principle), unless it has a sufficient reason for being thus and not otherwise. As stated, analysis can never uncover what this reason may be in any concrete case, because it would then have to do with actually existing monads, which have an infinite variety of perceptions, or with bodies,which are actual infinities of monads, and so are infinitely divisible. In Leibniz's own words, such an analysis is like the case of surd ratios in which "the reduction involves an infinite process and yet approaches a common measure so that a definite but unending series is obtained, so that contingent truths require an infinite analysis, which God alone can accomplish" (in: On Necessary and Contingent Truths). Only God can determine what that sufficient reason is because only he has such clear perceptions that by a mere *act of intuition* he has a complete and adequate *existential* knowledge of the whole of the universe, and hence also in any part. Thus truths of reason and truths of fact *truly exist*; Leibniz is no nominalist! But because of man's finite nature, man is able to have only a *symbolical* knowledge of these truths. Only in the case of numbers does man have something of an intuition that at the same time gives adequate knowledge of the truth. In every other case, however, man must seek an inadequate way to come to that knowledge,e.g., such as in algebra.This brings us to Leibniz's contribution to symbolic logic.

Leibniz agreed with Descartes that knowledge could be clear and distinct or obscure and confused. However, Leibniz differed from Descartes in that, over and above this criterion, one had to ask how that which one strove to attain knowledge of was *possible.* This question was called for, for instance, from the side of mathematics. Leibniz had found that the concept of the largest number was selfcontradictory, hence impossible. Also in other areas, such as physics and theology, he had found impossible concepts, which, according to the criteria of clearness and distinctness, seemed at first to be true. To this end, Leibniz tried to set up a calculus or a method of combining symbols which would ultimately be able to account for, i.e., show the possibility of, every truth.

His approach was via concrete representations which, in suitable symbolism, is the 'thread of Ariadne' which leads the mind. His program was first of all to devise a method of so forming and arranging characters and signs, that they represented thoughts, that is to say, that they were related to each other as the corresponding thoughts. Once in possession of this *characteristica universalis*, a method of symbolic reasoning or calculation – *calculus ratiocinator* – was needed. This calculus had to include every valid form of reasoning with the same degree of certainty and exactness as an algebraic calculation. Then there had to be an *ars combinatoria*, a way of defining new symbols from old ones.

Leibniz's contribution to symbolic logic lies herein, that he was the first to formulate what would be necessary for a complete logical system. It was a *programme*, and much beyond its formulation Leibniz did not go.

The meaning of this symbolic logic, however, must be seen in connection with man's finite nature. In principle, every truth has real existence, though only God can know them fully on account of his unblemished perception (intuition). It is easy to see that when theological speculation became less popular, this realistic side of Leibniz's thought underwent reinterpretation while the formalistic/symbolical logistic side came to the foreground. It is in this vein that Bertrand Russell learned from Leibniz and carried on the logistic programme.

II.2.3.4. Kant (1724-1804)

It was Kant who tried to overcome the differences between the Newtonian conception of the concepts of natural science (and hence of mathematics) and the Leibnizian one. He proposed a third way of viewing these fundamental concepts. Furthermore, Kant tried to bridge the difference between the view of reason as exclusively geared to the natural sciences and mathematics – that which he called pure reason – and the view of reason as engaged in the practice of life – practical reason. He did this by introducing a 'transcendental' problematics, i.e., an investigation into the *necessary* (*a priori*) *conditions* which make both types of reason *possible*. Kant's view of reason became a critical one, because it had to give a critique of both pure and practical reason. Since Kant saw in mathematics (and mathematical physics) the prototype of a general critique of human knowledge (metaphysics), it interests us all the more how Kant saw mathematics, both in the nature of its concepts and its judgments (propositions). Needless to say, only the solution of the mathematical problem will be our concern, not the metaphysical one.

Kant proceeds to this method by first making a distinction in types of *a priori* judgments. Ever since Leibniz, all *a priori* judgments were considered analytical, i.e., in compliance with the law of contradiction. The empiricists had another type of judgment, namely, synthetical ones, whereby experience taught the mind to relate this concept to that one, simply because they always occur together. There was no necessity in this connection. Kant now saw that there were also *necessary* synthetical judgments, i.e., *a priori* judgments which still 'say something' in the sense that they were not reducible to an identity (hence could not fall under the law of contradiction) but which nevertheless bind concepts in a necessary fashion. To gain clarity on this problem, Kant turned to mathematics, a domain where he saw this to be the case.

Thus, mathematics is seen by Kant as based on synthetic *a priori* judgments. Kant did not mean to deny the validity of the law of contradiction in mathematics. But he saw the latter to apply only in the way one drew conclusions and deductions from initial axioms. Also, judgments such as 'the whole is equal to the sum of its parts' were seen to be analytic *a priori*. The initial axioms, however, he saw as synthetic in nature. He accounted for this as follows:

Take the judgment '5 + 7 = 12'. This is synthetic, because, by taking the sum of the concept '5' and the concept '7', it is not evident which number will exactly coincide with this sum. One of the two numbers must be represented in an intuition of points, fingers, or whatever you will, and then these units in intuition must be successively added to the concept of the other number. In this manner, it is *calculated* what number is identical to the sum of the two other numbers. In an analogous fashion, the intuition is necessary in geometry, in order to see the necessary connection between the concepts 'straight line' and 'shortest line between two points'. For here, the concept 'straight line' is purely qualitative, according to Kant, whereas the second concept is quantitative and must accordingly be represented in an intuition before we are able to see the necessary character of the judgment: "the straight line is the shortest line between two points". The intuition plays an essential role here because "the question is not *what* we must join in thought to the given concept, but what we actually think together with and in it, though obscurely; and so it appears that the predicate belongs to this concept necessarily indeed, yet not directly but indirectly by means of an intuition which must be present" (Kant (2). 269).

For Kant, pure mathematics is the following:

> The essential and distinguishing feature of pure mathematical knowledge among other a priori knowledge is that it cannot at all proceed from concepts, but only by means of the construction of concepts. As therefore in its propositions it must proceed beyond the concept to that which its corresponding intuition contains, these propositions neither can, nor ought to, arise analytically, by dissection of the concept, but are all synthetical. ((2), 272; see also (1), Methodology, Chapter 1, Section 1)

Having explained the character of mathematical knowledge, Kant proceeds to ask the transcendental question: What are the necessary conditions which make mathematical knowledge (i.e., mathematical synthetic *a priori* judgments) possible? Kant seeks the answer by analysing the notion of intuition that he has found to be involved in the foundations of mathematics.

First of all, the intuition must be pure, i.e., *a priori*. An impure or empirical intuition is contingent because it contains sensations from the outer world. The pure intuition, in which all mathematical concepts can be exhibited or constructed, is the "clue to the first and highest condition of its possibility".

The pure intuition, being *a priori*, had to be prior to any actual intuiting of any object. But what is then intuited in this pure intuition? "In one way only can my intuition anticipate the actuality of the object, and be a cognition *a priori*, namely: if my intuition contains nothing but the *form* of sensibility antedating in my mind all the actual impressions through which I am affected by objects". (Kant (2), 282) The form of sensibility for outer impressions, for Kant, is space, and that for inner impressions, time.

Thus the intuitions, which pure mathematics lays at the foundations of all its cognitions and judgments, are space and time. Geometry is based on the pure intuition of space. Arithmetic achieves its concept of number by the successive addition of units in time. These are the pure and apodictic factors which make mathematics possible.

Because the mathematics of Kant's day knew no other geometry but Euclidean geometry, it is natural that Kant identified the spatial sensibility form with the geometry of Euclid. It is commonly thought that, with the appearance of non-Euclidean geometry, the *entire* Kantian notion of geometry as a system of synthetic *a priori* judgments was thereby contradicted. This is not correct. Not only does the precise Euclidean character of geometry have nothing to do with Kant's arguments for his view of geometry alone with space as an *a priori* sensibility form, but also later thinkers had little trouble re-interpreting Kant's ideas for different geometries. B. Russell, in his dissertation, re-interpreted Kant with respect to projective geometry, and later, Poincaré re-interpreted Kant with respect to topology. (Poincaré (4), p.44)

We end this discussion with a note on the status of the infinite in Kant's thinking. It can be said that Kant was one of the first to 'do in' a danger resulting in the liberal ways in which the 17th and 18th century thinkers had worked with the notion of (actual) infinity and (actual) infinitesimal. (Earlier, Berkeley had merely ridiculed these notions.) Kant did this by investigating the status of the infinite in ontological (cosmological) questions. In this respect also, Kant came to differ with both Newton and Leibniz. For both of these thinkers had given a *constitutive* meaning to the infinite. The substratum of the world, for Newton, was given by absolute structures of (infinite) space and (infinite) time. In Leibniz, each monad had an actual infinity of perceptions and each body was an actual infinity of monads. Kant found inherent antinomies in this constitutive sense of infinity (see his first and second antinomy of pure reason), which came down to the following: when the world, or anything in it, is considered as finite, the understanding is able to think it enlarged: when the world, or anything in it, is considered as actually infinite, the understanding cannot think it at all. In both cases, reason is incongruent with the world. For the finite is too small for reason and the (actual) infinite too big (cf. Kant (1), A 485 ff).

The solution Kant offered was to see the infinite, not in a constitutive sense, but in a regulative sense. He explains himself as follows:

> ... it is a principle of reason which serves as a *rule*, postulating what we ought to do in the regress, but not anticipating what is present in the object as it is in itself, prior to all regress. Accordingly I entitle it a regulative principle of reason, to distinguish it from the principle of the absolute totality of the series of conditions, viewed as actually present in the object ... which would be a constitutive cosmological principle. ((1), A 509)

With this shift of meaning, the notion of the infinite is taken out of ontology and transferred to epistemology. As will be seen, not all the schools of 20th century mathematics were happy with its shift.

II.3. TRANSITION TO THE PRESENT CENTURY

In Kant's work, the type of philosophising called transcendental *idealism* was introduced. This idealism, which dominated the academic world till Hegel's death (1831), was the final result of viewing reason as containing static *a priori* (mathematical) *concepts* and *judgments*. The rise of the experimental sciences – biology, psychology, thermodynamics – along with the developments of technology, brought a change in the philosophical world as to the nature of reason. Positivism arose. Reason came now to be seen as containing, *a priori,* the methods whereby scientific knowledge was to be obtained. This gave a kind of 'freedom' to mathematics in that numerical and spatial concepts were not necessarily seen to be *a priori.* Accordingly, along with the experimental work done at this time, an *empiristic* view of mathematics came to the foreground. Especially geometry was developed as an empirical science, e.g., by Gauss, Riemann, Pasch, Helmholtz, etc. With Helmholtz, for example, we see the attempt to base geometry on mechanics! Techniques for measurement became important, which, in a way, came to a peak with Riemann's idea of the metric in geometry as a function of mass. This lead Einstein to see the gravitational field as a metric field, which is clearly dependent on the mass at any point in space.

The proliferation of mathematics and logic in the 19th century are well-known. In metric geometry, Euclid's axioms were removed from their pedestal by the appearance of Lobatschewskian and Riemannian geometry. Also projective geometry came to be seen as an independent non-metrical geometry in its own right. Furthermore, in algebra came the discovery of group and ring structures. The group structure, via the theory of invariants, did much to account for the lay-out of the various geometries. Also logic was seen more from its mathematical side. In the spirit of Leibniz's *calculus ratiocinator*, Boole, De Morgan and Schröder developed an algebra of logic (a form of the logic of classes). In this proliferation, the relation between mathematics and logic became an important question.

In attempting to overcome the problem posed by that relation, we meet, for the first time, the direct interaction of previous philosophies of mathematics with foundational problems. Previous to this time, mathematicians had done their work without having consciously before them a philosophy of mathematics. (On the whole, the presuppositions of mathematics were of a realistic nature. But many mathematicians were not always sure what this meant.) Correspondingly, it had been, on the whole, *philosophers* who had formulated a philosophy of mathematics. Now with mathematicians turning to an investigation of the foundations of their work, the philosophical standpoint of the investigator could no longer be ignored. It is not very surprising that the philosophical positions taken were, roughly speaking, either a Leibnizian-logistic-realistic one or a Cartesian-Kantian-intuitionistic-constructivistic (idealistic) one.

Frege and Peano, via their work in symbolic logic, defended the logistic thesis that mathematics is but a chapter in the book of logic. Frege worked this out for arithmetic, and Peano tried to encompass the whole body of mathematics in a logical symbolism. But it was not until Russell came and wrote, in collaboration with Whitehead, the *Principia Mathematica* that the full scope of the logistic thesis was developed and seen in its full power – and weakness!

Another attempt at unification was started via the constructive method. Kronecker came to a number theory that developed all the types of numbers, e.g. rationals, reals, complex, etc., from the natural numbers, which, in turn, were *a priori* and intuitively recognized. The heart of this attempt he captured in the epigram: "God made the natural numbers, everything else is the work of man". Borel, Lebesque, and especially Poincaré continued in this line and became known as *intuitionists*. When Brouwer came on the scene, the intuitionistic program was radicalized to a more strict (finite) constructivism. His views, along with those of his followers, are called *neo-intuitionism*.

At this point, within both the context of our story and the time in history being described, Cantor (1845-1918) must be mentioned. From out of his work in analysis, especially Fourier analysis, he came to a profound study of the notion of infinity, based on a very general definition of a set. This was the following: "Unter einer 'Menge' verstehen wir jede Zusammenfassung *M* von bestimmten wohlunterschiedenen Objekten unserer Anschauung oder unseres Denkens (welche die 'Elemente' von *M* genannt werden) zu einem Ganzen" (Cantor, cf.: Beiträge zur Begründung, par. 1).

Historically he is important, because it was within the consequences of this theory that the famous paradoxes of the 'set of all cardinal numbers', the 'set of all ordinal numbers', etc., were found. (Upon hearing this, Russell then found his logical paradox on the set of all normal sets.) Now, as never before, both mathematics and logic were faced with grave problems in their development. Of equal importance, if not more, is the systematic side of Cantor's work. Whereas the various views of the infinite had been a bone of contention for the *philosophers* – recall the difference between Kant and Leibniz on this point – and the antinomics of reason had been philosophical ones, now a more or less identical problematics sat square in mathematics and logic itself. The notorious notion which caused this transformation was precisely the notion of a *set*. Set theory was the final and most general foundation of *analysis* and of the notion of *infinity*. The problematics as to the status of the infinite were now, mutatis mutandis, the problems as to the status of the notion of a set.

With this 'bird's eye view' of the problems, we turn to a more detailed look at the tenets of the more predominant spokesmen of the various foundational directions.

II.4. DIRECTIONS IN THE 20th CENTURY PHILOSOPHY OF MATHEMATICS

II.4.1. Logicism

The logicistic thesis has been succinctly summarised by Carnap (p. 31) as follows:

(i) The *concepts* of mathematics can be derived from logical concepts through explicit definitions;

(ii) The *theorems* of mathematics can be derived from logical axioms through purely logical deduction.

We see immediately how this differs from empiricism and intuitionism. Empiricism considers mathematical entities to be dependent on physical or psychical lawfulness, and views these sciences, as a whole, to be *a posteriori.* Intuitionism defends the view that mathematical concepts are constructed, and that the science, as a whole, is synthetic. In logicism, mathematical concepts are explicitly defined and its theorems are analytic. Since logic is *a priori*, mathematics, as its derivative, is necessarily so.

To make headway in this program, logic had to be made more powerful than it was as Aristotelian logic. The syntax (Leibniz's 'characteristica universalis'), the calculus of reasoning ('calculus ratiocinator') and the definitions ('ars combinatoria') had to be expanded. It was the calculus ratiocinator that developed first in the form of the logical algebras of Boole, De Morgan and Schröder. The logic of relations was developed by De Morgan and Peirce and later much more by Russell and Whitehead. The propositional logic, the actual cornerstone of the logic of logicism was developed by Peirce, Frege and Russell.

Russell and Whitehead, in their *Principia Mathematica*, brought the program to its nearest completion. The only main branch of mathematics lacking was the logistic view of geometry. Of this Russell discloses, on the occasion of Whitehead's death (1947): "Whitehead was to have written a fourth volume (of *Principia Mathematica*) on geometry, which would have been entirely his work. A good deal of this was done, and I hope still exists. But his interest in philosophy led him to think other work more important. He proposed to treat a space as the field of a single triadic, tetradic, or pentadic relation, a treatment to which, he said, he had been led by reading Veblen". (Quoted from Wilder, p.242)

II.4.1.1. Frege (1848-1925)

Frege makes a fundamental distinction between what he calls 'Sinn' and 'Bedeutung' of complete symbols. Sinn refers to the manner in which an object (Gegenstand) is expressed or referred to, while Bedeutung of a complete symbol is the actual Gegenstand. Thus '2 + 2' and '4' have the same Bedeutung but each is distinct in Sinn. Any sentence is also a complete symbol. Its Sinn is identical with the thought it contains; its Bedeutung is its *truth-value*. Frege's assertion sign '⊢' is built up from these distinctions:

'*a*': the Sinn of a sentence, symbolised by '*a*', independent of its truth-value.

'-*a*': the Bedeutung of the sentence. This can only mean 'true' or 'false'.

'⊢*a*': the sentence when true. It means that '*a*' is *asserted*. It expresses an acknowledgement of truth.

Frege also symbolised the logical operations: negation, implication, disjunction and conjunction. He then introduced his logical foundation of the number concept, which is roughly the following.

His starting point is the simple fact that the cardinality of two sets can be relatively expressed without the use of the number concept. Now a set, for Frege, is the extension of a concept, denoted by F or G. We will simply speak of the concepts F and G, meaning thereby the extension of these concepts.

Now F and G are equipollent if there is a 1-1 correspondence R between the objects falling under F and those under G. The concept ‘cardinality of F’ means the same as ‘equipollent with F’. A small letter n will denote the cardinality of a concept if and only if there is a concept F such that the concept n means the same as ‘equipollent with F’.

This says what cardinality *is*, but does it exist? To show this, form the concept ‘not identical with itself’. This means that a concept has no extension and we denote its cardinality with 0.

Now define ‘identical with 0’. The extension of this concept only encompasses 0 and its cardinality is 1.

Now define ‘identical with 0 or 1’. The extension of this concept encompasses 0 and 1 and its cardinality is 2.

It is easy to see that every natural number can be acquired in this manner from logical concepts. *Any* natural number may be defined, but that does not give us the general concept ‘natural number’.

To define the concept ‘natural number’, one proceeds as follows: *Let ‘F’ be a concept and let ‘a’ be an object not falling under F. Define F as ‘everything under F or identical with a’*. If n is the cardinality of F, then n' is the cardinality of F'. Now define a function h from cardinality to cardinality:

if from F one can define F' by means of an object a not under F, then from the cardinality n of F we map the cardinality n' of F', i.e., $h(n) = n'$; if no object 'a' exists, then $h(n) = n$. As long as objects not falling under the concepts dealt with exist, one can build up a chain of h's starting from 0. The concept 'natural number', for Frege, is the same as the concept 'belonging to the h-chain of 0'. From this definition, the usual description of natural numbers (Peano axioms) can be derived. And for just that reason we need go no further.

Frege's fight against the psychologism (including empiricism) of his day made him eager to give a non-psychological interpretation of his notion of assertion, and hence of his use of the notion of concept or set. He thereby introduced a *platonic realism into his logicism* that *was to guarantee that concepts were not constructions* or figments of the mind but, rather, were non-arbitrary entities. But because concepts are 'non-physical' entities, they could have no actual existence.

Frege's solution is expressed in his dictum: "Ich erkenne ein Gebiet des Objektiven, Nichtwirklichen an". (In this there is the influence of Lotze; cf. Beth (2), p.353.)

The novelty introduced by Frege for the interpretation of the number concept lies in the fact that a number is a *property* of these objective, non-actual concepts. One sees immediately how opposed this is to *empiricism and intuitionism*: the former makes number an attribute of *actual* things; the latter makes number subjective, being a construction of the mind. (We shall also call this latter position *idealism*.)

II.4.1.2. Russell (1872-1970)

We cannot go into all the facets of Russell's development. We mention only the turns he gave to the logicistic programme in its attempt to subjugate the logical antinomies inherent in Frege's approach, as discovered by Russell.

The problematic character of the paradox, which will be mentioned shortly, can be directly attributed to Frege's realism: his belief that there were entities in a realm accessible only for thought. This belief led him to *substantialize* and see as entity everything the mind *thought*. A set, being an abstract entity, had at the same time an existence of its own. It could be referred to or asserted without reference to anything else. It was a 'complete symbol'. That this leads to difficulties is best described by 'Russell's Paradox'. Consider the set of all normal sets: i.e., all those sets that do not contain themselves as elements will be called abnormal. Evidently, every set is either normal or abnormal. Consider now the set of all normal sets. Is this set normal or abnormal? If it is normal then it cannot contain itself as an element. But this makes it eligible for membership in the set of all normal sets; hence, it does contain itself as an element. Therefore, it must be abnormal. By similar reasoning, if the set of all normal sets is abnormal, it must be normal. Either way a contradiction results. The problem was, as Russell saw it, that one went too readily from referring to a set to substantializing it, i.e., to making it capable of serving as an element for another set. Or, in reference to Cantor's definition of a set, it turned out to be not at all clear in what sense the "Zusammenfassung van bestimmten wohlunterschiedenen Objekten" was indeed a whole, a new entity, a new truth.

Russell's ultimate solution to this perplexing state of affairs involved stating the minimum restrictions necessary to avoid the paradoxes. Russell was *not* prepared to give up the realism. The remedy for mathematical logic that he presented in *Principia Mathematica* was his *theory of types*. The theory of types is first of all an hierarchical arrangement of sets. The sets are arranged in types. Type 0, the first type, contains only individuals. Type 1 contains properties of these individuals, hence also sets which have their elements in type 0. Type 2 contains properties of the properties of type 1, or sets whose elements are sets of type 1. This can be carried on as far as one pleases. This theory made impossible formulations that allowed a set to contain itself as an element. For now the set and its elements belong to different types, thereby precluding that a set can be its own element. Only those sets can become elements of another set if the first are of type k and the other $k + 1$.

This theory has, as a consequence, that the notion of a set is also an incomplete symbol. For something beyond the set itself is needed to determine whether the set referred to indeed denotes a meaningful, i.e., true or false, entity. For Russell, a set – or class, as he usually called it – is contextually defined, whereas Frege saw in each set an entity, complete in itself. For this reason, Russell termed his theory a 'no class theory'. Only of complete propositions can one meaningfully speak of truth or falsehood. Russell did not deny, at this stage of his development, that there was mathematical truth; he only denied its meaningfulness in reference to incomplete notions.

With this proliferation of types, it was imperative that some kind of reduction would be legitimate. For instance, one consequence of the theory is that each type would demand its own number system. Things which seemed to be identical were now no longer so, should they end up in different types. The 'axiom of reducibility' that Russell incorporated was, as he says in *Principia Mathematica*, a version of Leibniz's identity of indiscernibles. If the extension of two notions in different types is the same, then it must be legitimate to consider these notions 'formally equivalent'. The 'axiom of reducibility' made this possible. Its agreement with Leibniz's principle lay herein, that, if two things had exactly the same predicates, they were then indiscernible.

Tarski has formulated the axiom as follows: Let $U(x_k)$ be a formula of class x_k in type k. Assume that U does not contain variables that run through type $k+1$. Then the axiom says: there exists an x_{k+1} such that, for all x_k $(U(x_k) \leftrightarrow x_{k+1}(x_k))$, where $x_{k+1}(x_k)$ means that x_k is an element of x_{k+1}. This says that on the $(k+1)^{st}$ level, one can determine what the U's on level k will be like.

This principle, though a main bone of contention for other logicists, had for Russell an important place in his philosophy. It was an application of 'Occam's Razor' (do not multiply entities beyond necessity), which Russell formulated as: "Wherever possible, substitute constructions out of known entities for inferences to unknown entities". This gave, among other things, a different interpretation of the logicistic definition of number. Whereas Frege saw a number as an attribute of a concept, i.e., inferring from equipollent concepts that the entity n must exist (cf. Angelelli), Russell now equated the number n with the class of all those classes that had that cardinality. Another example of this principle is seen in the definition of a real number. Whereas Cantor defined a real number as the limit of a convergent sequence of rational numbers – thereby inferring the existence of an unknown entity from known ones, Dedekind identified a real number, via the idea of a cut, with the lower set of rationals formed by the cut. Needless to say, Russell chose the latter.

The full account of the logicistic thesis that mathematics is derived from logic required, besides the axioms of logic, an axiom of infinity, the axiom of choice, and the above mentioned axiom of reducibility. These latter axioms are of existential import, in contradiction to the logical axioms which only lay claim to what is possible. The objection has been raised in how far one can claim that mathematics is now indeed derived from logic. For the existential axioms are, in fact, mathematical axioms. In any case, they were introduced to make the further derivation of mathematics possible. The continued investigations of these questions within logicism, e.g., by Carnap and Quine, have led to different directions in the development of the program. But no one has been able to bring to a completion the logicistic programme, as envisaged by Leibniz, Frege and Russell, which is at the same time free from inherent difficulties, or unnatural traits.

The logicistic-realistic position did lead to further developments in philosophy. A case in point is Whitehead's philosophy. From the logic of relations, as developed in the *Principia Mathematica*, he came to a view of reality in which relations played a fundamental role. But since this is a case of mathematical logic influencing philosophy and not *vice versa*, we will not develop this train of thought further.

Another development, more on the epistemological side of this school, arose with Wittgenstein's convincing Russell that mathematical logic is a grand tautology. This meant that 'truth' and 'meaning' played no role in mathematics at all. It was this step that brought former logicists to consider their work more from a linguistic side. The step is somewhat analogous to Russell's treatment of the class notion as a nominal notion as opposed to Frege's realistic notion of it. This shift from logic to language, from less realism to more nominalism, is what characterizes the 'neo-positivistic' movement, and it, accordingly, has investigated new semantical and syntactical aspects of theories in their relation to life.

II.4.2. Intuitionism

As opposite – almost completely opposite – to logicism stands intuitionism. Here we use the name 'intuitionism' to stand for every view according to which neither logicism nor formalism are able to grasp the essential feature of spontaneous mathematical thought. Intuitionists, in this broad sense, thus attach less value and importance to the role of formal logic in mathematics. This does not necessarily mean that they reject the axiomatisation and formalisation of mathematical disciplines. The most vocal and authoritative spokesman for the initial development of this movement in the philosophy of mathematics was Poincaré. Accordingly, we shall first look at his position before making references to further developments in this movement by the Dutch mathematician, Brouwer, and others.

II.4.2.1. Poincaré (1854-1912)

Poincaré, besides being a brilliant mathematician, showed great interest in the foundations of his work. On all counts, he was the most serious initial opponent logicism was confronted with. And because of his authority, for many decades after his death France made next to no contribution to the development of mathematical logic. All the more reason to observe this man's thoughts closer.

It was Poincaré's firm opinion that mathematics was not a derivative of logic. This he defended for the two main domains of mathematics: number and space. But first, what is logic?

Logic, for Poincaré, is only at home in the finite and its main task is to classify. In syllogistic reasoning, in which one has placed the things reasoned about in fixed classes, one draws conclusions on the basis of the relations between these classes. Any attempt at importing an 'axiom of infinity' into logic is treason: one cannot reason about objects which cannot be defined in a finite number of words (Poincaré (4), p. 60). It is only in mathematics that one can refer to the (potential) infinite, because mathematical reasoning may be reason by recursion. The principle of mathematical induction, as an eminent means of reasoning by recursion, is distinct from logic precisely in that this principle deals with a never ending *process.* The basis of this process lies in the *intuition.*

Intuition and logic need each other. Whereas logic can only analyse and *combine,* the intuition provides the *unity or harmony* of the demonstration. Logic gives rigor and certainty but is tautologous. Intuition provides the initial givens, it invents and unifies (Poincaré (2), p. 18). Logic separates a complexity into its elements and studies its relations. Logic may also investigate the possible recombinations of these elements into new aggregates. But the intuition discerns which combinations are relevant, by chosing just those combinations which show an analogy to other facts. Thus, construction is only a necessary factor for mathematical advance, and not a sufficient one. (Poincaré (1), p. 15)

Poincaré distinguishes two kinds of intuition relevant for mathematics: the intuition of pure number and the intuition via the senses and imagination. The former is indubitable and is 'above' the senses and gives us the basis for arithemetic; the second is less certain and gives us the basis for geometry. Poincaré emphasizes the essential difference between these intuitions: "they have not the same object and seem to call into play two different faculties of our soul; one would think of two search-lights directed upon two worlds, strangers to one another". (Poincaré (2), p. 25) We now say something about each.

The intuition of pure number borrows nothing from the outside world. It is *a priori.* This *a priori* character of the natural numbers is not what interests Poincaré. (We shall see later what his criterion is for mathematical existence). Of greater consequence is the nature of the reasoning one performs when working with numbers. Poincaré readily admits that any numerical equation, e.g., $2 + 2 = 4$, is analytic (contrary to Kant) and hence, logical. But in order to advance from arithmetic to algebra one must prove properties that are to hold for *all* the natural numbers. So as not to introduce the notion of actual infinity, that property can only be proven to hold if, by mathematical induction, it is shown to hold for each number. Thus, if shown to hold for *any* number, it will hold for all. Now reasoning by mathematical induction is a process of repeating the same operation with each number, and, because there is no (sufficient) reason for stopping at any number, one realizes that reasoning by recurrence is "the only instrument which enables us to pass from the finite to the infinite" (Poincaré (1), p. 11). Since logic can only deal with what is finite, this principle of mathematical induction is irreducible to the law of contradiction. This mathematical principle is *a priori* and not analytical. It is of a *synthetic* a priori character. But it is not a judgment, as Kant had it, it is an intuition. Its irresistible nature lies herein, that "it is only the affirmation of the power of the mind which knows it can conceive of the indefinite repetition of the same act, when the act is once possible. The mind has a direct intuition of this power . . . Mathematical induction is necessarily imposed on us, because it is only the affirmation of a property of the mind itself". (Poincaré (1), p. 13)

The science of space, i.e., geometry, must have an outside world to make the very rise of geometry possible. Is Poincaré an empiricist in geometry? Not entirely.

Via the intuition of the senses, Poincaré explains how we

come to our notion of an empirical *representational space.* This empirical space is representational, for it is that which is represented in our sensations. We have sensations of touch, sight and muscular effort. We form a notion of space – or extension – from the combination of these sensations by the fact that we must perform bodily motions to reach – or imagine that we reach – the objects represented in our sensations. We form a geometry in this representational space by investigating the way these sensations succeed one another. Hence, without external bodies in motion, we would not be able to form an empirical notion of space and geometry. In forming these empirical notions, our bodies function as the axes of reference, the source of coordination.

We go from this representational space to a *mathematical space* by *reasoning* about the objects in representational space as if they were given in a mathematical space (or continuum). The external bodies become rigid figures and the motion of these bodies, i.e., the succession of sensations, is idealised into the mathematical concept of a *group* of motions. But, many different laws of succession of sensations are possible; accordingly, there is no one geometry that 'fits' the world. Everything that we account for in the world by Euclidean geometry may also be accounted for by non-Euclidean geometry. This also gives no special status to the axioms of any one geometry. The axioms are not synthetic *a priori* judgments, but are *disguised definitions* of how we idealise our representational space into a mathematical one. Because many geometries are possible for use in the world, it is of mere *conventional* character which geometry may be used. Simplicity and elegance are the usual norms.

Poincaré proceeds to show how our notions of point, line, plane, dimension, etc. are all idealisations of sensations. However, a complete empiricism is made impossible. Poincaré sees that ultimately there must be a (Kantian) *form of sensibility* which will make possible the very ordering of sensations into a space, and a *form of understanding* which will make possible the very notion of a geometry. The latter is the notion of a group, the general concept of which 'pre-exists in our mind, at least potentially'. (Poincaré (1), p. 70)

Of the form of sensibility, Poincaré says:

> . . . there is in all of us an intuitive notion of the continuum of any number of dimension whatever because we possess the capacity to construct a physical and mathematical continuum; and that this capacity exists in us before any experience because, without it, experience properly speaking would be impossible and would be reduced to brute sensations, unsuitable for any organization; and because this intuition is merely the awareness that we possess this faculty. And yet this faculty could be used in different ways; it would enable us to construct a space of four just as well as a space of three dimensions. It is the external world, it is experience which induces us to make use of it in one sense rather than in the other. (Poincaré (4), p. 44)

With this summary of Poincaré's philosophy of mathematics, we briefly mention his polemics with logicism.

Since Poincaré rejects radically anything involving an actual infinity, he was by no means inclined to any realism with respect to mathematical existence. Poincaré emphasizes that the only criterion necessary for mathematical existence is that the mathematical concepts and the system in which it appears be *free from contradiction.* This notion had played the important role of establishing the non-Euclidean geometries. As is well known, at the end of the 19th century, it was found that there was a complete isomorphism between the structure of Euclidean and non-Euclidean geometry. This meant that, should a contradiction arise in the one, the other would show one also. Now Hilbert, in his *Grundlagen der Geometrie* (1899), had shown that a similar isomorphism existed between Euclidean geometry and the real number system. Since the latter could be constructed from the natural numbers (along with an adequate set theory), if ordinary arithmetic is free from contradiction, then the whole of mathematics has a good right to exist. Any logicist, of course, would agree with this. But Poincaré goes on to say that arithmetic cannot be proven to be free from contradiction. For, to show that elementary arithmetic is free from contradiction, we would need to show that an infinite number of theorems is free from contradiction. And this could only be proven, if, *by mathematical induction,* we progressively show the system to be free from it. However, since one needs the very principle which is, at the same time, included *in* arithmetic as one of its axioms, to establish this exemption from contradiction would incur a *petitio principii.* Though logicism meant to base even arithmetic on logical primitives, and thereby give a firm basis to it, Poincaré sees in this a begging of the question, in that the axiom of reducibility is but a disguised form of the principle of mathematical induction.

The first sin of logicism is that it has not shown its axioms to be free from contradiction, – and, indeed, it could not have done so on account of the above-mentioned *petitio principii.* Only by calling on the intuition can the true basis of mathematics be seen.

The second sin of logicism is the supposition that, once the initial primitive logical notions are given, mathematics will roll out without further appeal to an intuition. But, for Poincaré, each demonstration must have a goal which gives unity to the demonstration. And this unity, this 'soul' of the demonstration, is also a matter of intuition.

The third sin of logicism is its realism. When logic is expanded to infinite collections, the usual rules of formal logic, i.e., the study of the properties common to all classification (Poincaré (4), p. 45), cannot be used unrevised. For with infinite collections, one must be aware that the classification does not change upon introducing new elements. This will be the case when each element of a collection is defined in terms of that collection. Such a classification is *impredicative*. For, with each introduction of a new element, the classification must be modified on account of the fact that the whole is changed. Poincaré will acknowledge only those classifications which are predicative. The impredicative classification results in the vicious circle type of reasoning that brings forth the antinomies in set theory and logic. Poincaré ((4), p. 63) suggests that the following rules should guide mathematics and logic:

(i) Never consider any objects but those capable of being defined in a finite number of words;

(ii) Never lose sight of the fact that every proposition concerning infinity must be the translation, the precise statement of propositions concerning the finite;

(iii) Avoid non-predicative classifications and definitions.

Poincaré's concern for always proceeding from the finite lies in the fact that the human mind (the intuition) is present in all mathematical work. Realists deny this, and imagine themselves dealing with a realm disjoint from the human subjectivity. They proceed from the infinite to the finite, and thereby make possible impredicative definitions and classifications. As this difference between realism and idealism (as he called his own position) lies in the *philosophical* foundations of mathematics, Poincaré was pessimistic of ever seeing this antithesis resolved completely.

On the notion of a set, the big trouble-maker in all this, Poincaré says the following ((4), p. 61):

If we consider a set and we wish to define the different elements in it, this definition can be broken down naturally into two parts; the first part of the definition, common to all the elements of the set, will teach us to distinguish them from the elements which are alien to this set; this will be the definition of the set; the second part will teach us to distinguish the different elements of the set from one another.

Each of these two parts will be made up of a finite number of words.

And for a set to have any meaning at all, *both* of these parts must be explicitly expressed.

II.4.2.2. Brouwer (1881-1966)

As with logicism, our concern will center on the philosophical background and implications of Brouwer's work, not that of his detailed intuitionistic mathematics and logic.

Brouwer's position, and that of his followers, is best referred to as neo-intuitionism on account of the fact that he radicalized the intuitionism of Poincaré at several crucial points. First of all, that of the very nature of mathematics. Whereas Poincaré kept mathematics close to, and partly in contact with, physics, Brouwer sees mathematics in a much broader human cultural context. Science, in general, is seen more from its social implications:

By science we mean the systematic cataloguing by means of laws of nature of causal consequences of phenomena, i.e., sequences of phenomena which for individual or social purposes it is convenient to consider as repeating themselves identically, – and more particularly of such causal sequences as are of importance in social relations (Quoted from: Benacerraf and Putnam, p. 66)

The humanizing role of mathematics, was brought forward already in one of Brouwer's theses, defended at his doctoral promotion:

De verstandhouding der menschen berust op het bouwen van gemeenschappelijke wiskundige systemen, en het verbinden aan eenzelfde element van zulk een systeem van een levenselement voor elk der individuen (stelling VIII, transl.: The understanding relationship between people proceeds from building common mathematical systems, and from assigning to the same element of such a system a vital element, for every individual).

With this basic societal role that mathematics must fulfil, it is natural that Brouwer should seek the basis of mathematics in the human subjectivity. This brings us to a second point of difference with Poincaré. Whereas the latter let the outer world play a large role in the genesis of the continuum concept and only let the number concept come from within, Brouwer wishes to have only *one* origin for mathematics. This origin must be the beginning of both number and the continuum. He does this by rejecting the Kantian-Poincaré notion of a spatial form of sensibility for sensations from the outer world, and laying all the emphasis on the inner form of sensibility, namely, the *experience of time*. If mathematics is to be pure, this inner experience must be freed of all passage of sensations present in our psyche. What results is a *purely intellectual formal notion of change in time*. This basis guarantees that mathematics will be completely *a priori* and completely free.

This purely abstract notion of change is the basis for both the ordinal numbers and the (first order) continuum. For, in becoming conscious of change, the *one* ego intuits itself in change (or variety, plurality). The becoming aware of change presupposes a twoness – for becoming aware of the ego in change, means to be aware of it as it was 'then' and how it is 'now' – in one and the same ego. This *intuition* of the *bare two-oneness* is what Brouwer calls the *basal intuition* (oer-intuitie). By repeating this basal act of intuition – "one of the elements of the two-oneness may be thought of as a new two-oneness" (Benac. and Putnam, p. 69) – any ordinal number may be counted off, or *constructed*. The natural ordinal numbers are the pure constructs of the intellectual intuiting mind. The linear (first order) continuum is constructed by repeatedly interspersing new units 'between' the given twoness. In this manner the linear continuum is constructed. As there is no last number, so there is always another 'between'. Because of this common origin, Brouwer says:

Waar dus in die oer-intuitie continu en discreet als onafscheidelijke complementen optreden, beide gelijkgerechtigd en even duidelijk, is het uitgesloten, zich van een van beide als oorspronkelijke entiteit vrij te houden, en dat dan uit het op zichzelf gestelde andere op te bouwen; immers het is al onmogelijk, dat andere op zichzelf te stellen (Brouwer (1), p. 8, transl.: As in the basal intuition the continuous and the discrete occur as inseparable complements, both with the same right and with the same clarity, it is therefore impossible to separate oneself from one of these as a basal entity, and build it up from the other one as an entity in its own right; just because it is impossible to separate off the other entity).

By making mental construction both a necessary and sufficient condition for the pure mathematical act – which constitutes a third difference with Poincaré – Brouwer would not bring in logic as a necessary complement to intuition. In fact, every human act, in that it is an act enacted in time, must involve the basal intuition of mathematics. Accordingly, logic and language are, at best, a help and a medium to making a mathematical act clear to others. But in no way do logic or language supplement mathematics proper. Hence, also, there is no 'truth' apart from what the human mind mathematically constructs. This had devastating consequences. For the logical law of excluded middle (the 'principium exclusi tertii') was now deprived of much of its service. By coupling mathematical existence with mathematical construction, it was no longer legitimate to place an object in one of two mutually exclusive categories if there was no means of constructing the object. The law of excluded middle had always played a vital role in the so-called 'existence' theorems, where, often, by a process of elimination, the existence of an object could be established entirely on the basis of the said law. It is obvious that much of higher mathematics – e.g., that part based on Cantor's transfinite cardinals and ordinals – was branded as meaningless by Brouwer. To make true his position, started already with his dissertation, much of Brouwer's life's work was directed to rebuilding mathematics from this constructivistic basis. To catch a glimpse of this reconstruction, see A. Heyting: *Intuitionism, An Introduction*.

II.4.2.3. The 'Bourbaki' group

The men of the 'Bourbaki' group, a number of French mathematicians who name themselves after a general of a previous century, perform their work much in the spirit of Poincaré.

We mention this group under the heading 'intuitionism', not because the Bourbaki consider themselves intuitionists, but mainly because they share with intuitionism, generally speaking, a concern for the natural function of thought in the mathematical science.

A main driving force of this group is the problem of the unity of mathematics. At the beginning of the 20th century, mathematics was a series of disciplines, founded on particular notions, delimited with precision and interconnected in many ways, thereby permitting one area to fructify the other. Continued investigation has shown the existence of a central core, which bridges the many diverse parts. The essence of this unity came to be seen in the progressive systematization of relations existing between the diverse mathematical theories and is today (for the Bourbaki group at least) known under the name of 'axiomatic method'.

Now the phrase 'axiomatic method' is usually an application of formal logic, or deductive reasoning. But this the Bourbaki group considers to be nothing other than an external form which the mathematician gives to his thoughts. It is there for communication and it is useful to have its vocabulary and syntax clearly analysed. But it is still only the surface of mathematics and the least interesting aspect of it at that.

According to the Bourbaki mathematicians, axiomatics catches what formal logic misses, namely, the profound *intelligibility* of mathematics. Where a superficial observer will see two entirely distinct theories, the mathematical genius may suddenly see a distinct analogy between them in their structure. And it is the task of the axiomatic method to search for such common structures in externally differing theories.

So far, all the known structures have been reduced to three fundamental 'mother' structures. These are:

(i) algebraic structures: i.e., structures which involve a law of composition, such as groups and rings;

(ii) (partial) order structures: i.e., structures defined by a relation between two elements;

(iii) topological structures: i.e., abstract formulations of the intuitive notions of neighborhood, limit, continuity.

From these structures, by combinations, all the particular disciplines of mathematics can be accounted for. These structures are the tools of the mathematician.

In mathematical research, the intuition – not the ordinary sense intuition – by a direct divination, grasps mathematical entities previous to all reasoning. And these entities, because of their frequent occurrence, have the same right of existence as beings in the outer world. The intuited structures carry with them their own proper language, full of particular intuited data. By means of an axiomatic analysis, the actual mathematical intuited structure is loosened from its surrounding language. The discovery of such a structure in the phenomena being studied, gives a sudden modulation to the researcher's thinking, in which the essential is distinguished from the peripheral. Less than ever is mathematics a game of purely mechanical formulas; more than ever is mathematics the field in which the intuition reigns in the genesis of discovery.

The fact that, in non-mathematical sciences, these fundamental structures are also being found, e.g., group structures in quantum mechanics, indicated that there can be no strict separation between the world of experiment and the world of mathematics. The mathematical structures are only part of the more general nature of things.

The meaning of 'intuition' in the Bourbaki group, as is clear, in no way excludes the importance of language and logic. Accordingly, we see, again, more of a complementation between intuition and logic. But, since the very subject matter intuited is caught in an axiomatic net, the Bourbaki group has given the impression of being extremely formalistic. That this is *only* an impression is evident when the task of the axiomatic method – as explained above – is made clear.

Summarizing, we may say that Bourbaki shares with Poincaré the insight that mathematics is not part of (or derived from) logic (or set theory), and that logic does not come under the mathematical sciences. Also the tendency of not making an artificial separation between pure and applied mathematics is inherited from Poincaré. The requirement that the mathematical sciences should be written in the perspectives of the 'mother structures', underscores once more the novelty of this approach. The reader may consult Bourbaki's original publication on their programme (cf. the bibliography) and also J. Piaget's interpretation of Bourbaki.

II.4.3. Formalism

Our story would not be complete without mentioning the third distinct direction in 20th century philosophy of mathematics: *formalism*. The formalism we have in mind was that defended by Hilbert (1862-1943). Since, here too, it is all too easy to say too much, we shall concentrate more on the philosophical side of this school than the mathematical. Also, we shall restrict our discussion to that of Hilbert's personal viewpoint.

Hilbert's formalism quite naturally developed out of his work in geometry. In this, as is mentioned earlier, he laid a complete set of axioms as a foundation for Euclidean geometry and then proved these axioms to be consistent by modeling them on the real number system. Geometry was now seen to be as consistent as (the operations in) the real number system. To put entire mathematics on a certain basis, free from contradiction, would involve the proof that the arithmetical operations are consistent. Logicism tried to meet it by a further modeling of arithmetic on logic. However, this went aground on account of the necessary acceptance of axioms which were more mathematical than logical in nature. Intuitionism contested that a formal proof of consistency could, in principle, not be given, and that intuitive reasoning, based on *evidence* and not on *axioms*, could be the only guarantee to mathematical truth.

Now Hilbert by no means tried to prove Brouwer wrong.

It is precisely in recognizing that the generally accepted mathematics are rather far removed from such statements as can claim real meaning and truth found on evidence, that Hilbert's own theory begins. But then he turns the tables on Brouwer. Instead of rejecting everything in mathematics that cannot claim real meaning found on intuitive evidence (construction), Hilbert proposes that we rid the whole body of mathematics of all meaning whatsoever. All mathematical statements are to be reduced to mere formulas, e.g., like the algebraic formula $a + b = b + a$. By this process, Hilbert was able to retain those mathematical statements which Brouwer had rejected on account of their inconstructibility, e.g., statements affecting the notion of infinity in some sense or other. These 'non-real' statements function in mathematics, according to Hilbert, like 'ideal propositions', which mathematics is full of. They are indispensable because they give a certain 'completeness' to the mathematical systems. And now that mathematics is to be drained of all meaning, the very distinction between real, evident statements and ideal ones vanishes. What remains to be shown is not that some formula is true or false – as had always been the case – but that the entire *system of formulas* is consistent. To do this, the system itself has to become the object investigated, and the character of that object is nothing but symbolical marks on paper.

Strange as it may seem, at this point Hilbert again shows that he has learned from Brouwer. But, as before, here also he turns the tables on him. For, Hilbert now maintains that one must become a strict intuitionist in investing this system of empty formulas for its consistency. The 'theory of proof', as he called this investigation, had to establish the *absolute* consistency of a mathematical system, and finitary methods are essential for this (cf. Hilbert). The insight of consistency must be attained by intuitive reasoning, for there is no other known consistent domain that the system can be modeled onto. But 'intuition' here does not mean the same thing as it did for Brouwer. What Hilbert is interested in, is the complete explication of the *rules* of deduction which one applies in deducing one formula from a set of other formulas, irrespective of content. And these rules are rules that must be *told* to others and communicated in *words*. Hence language – in contradistinction to Brouwer – forms the intuitive basis for the theory of proof (or metamathematics, as Hilbert also calls it). Language is parallel to the very technique of thinking.

Das Formelspiel, über das Brouwer so wegwerfend urteilt, hat ausser dem mathematischen Wert noch eine wichtige allgemeine philosophische Bedeutung. Dieses Formelspiel vollzieht sich nämlich nach gewissen bestimmten Regeln, in denen die *Technik unseres Denkens* zum Ausdruck kommt. Diese Regeln bilden ein abgeschlossenes System, das sich auffinden und endgültig angeben lässt. Die Grundidee meiner Beweistheorie ist nichts anderes, als die Tätigkeit unseres Verstandes zu beschreiben, ein Protokoll über die Regeln aufzunehmen, nach denen unser Denken tatsächlich verfährt. Das Denken geschieht eben parallel dem Sprechen und Schreiben, durch Bildung und Aneinanderreihung von Sätzen (emphasis not mine; quoted from Beth (3), p. 141).

Hilbert's type of intuitionism he himself describes as follows:

As a further precondition for using logical deduction and carrying out logical operations, something must be given in conception, viz., certain extralogical concrete objects which are intuited as directly experienced prior to all thinking. For logical deduction to be certain, we must be able to see every aspect of these objects, and their properties, differences, sequences, and contiguities must be given, together with the objects themselves, as something which cannot be reduced to something else and which requires no reduction. This is the basic philosophy which I find necessary not just for mathematics, but for all scientific thinking, understanding, and communicating. *The subject matter of mathematics is, in accordance with this theory, the concrete symbols themselves whose structure is immediately clear and recognizable* (emphasis mine; Hilbert, p. 142).

The novelty of Hilbert's approach is not to be denied. Hilbert himself hoped thereby "die Grundlagenfragen einfürallemal aus der Welt zu schaffen". With the actual mathematical statements reduced to empty formulas and the finite metamathematical reasoning immediately clear and evident, Hilbert thought he had made mathematics a science free of all presuppositions. So that he could say, full of confidence:

Schon jetzt möchte ich als Schlussergebnis die Behauptung aussprechen: die Mathematik ist eine voraussetzungslose Wissenschaft. Zu ihrer Begründung brauche ich weder den Lieben Gott, wie Kronecker, noch die Annahme einer besonderen aus das Prinzip der vollständigen Induktion abgestimmten Fähigkeit unseres Verstandes, wie Poincaré, noch die Brouwersche Urintuition und endlich auch nicht, wie Russell und Whitehead, Axiome der Unendlichkeit, Reduzierbarkeit oder der Vollständigkeit, die ja wirkliche inhaltliche und durch Beweise der Widerspruchfreiheit nicht kompensierbare Voraussetzungen sind. (Quoted from Beth (3), p. 141)

That this approach also proved to be 'too good to be true' is now a historical fact. Weyl, in his Hilbert obituary, describes this fact clearly and completely. We end this discussion of Hilbert's formalism with a rather long quote from that obituary:

The symbolism for the formalization of mathematics as well as the general layout and first steps of the proof of consistency are due to Hilbert himself. The program was further advanced by younger collaborators, P. Bernays, W. Ackermann, J. Von Neumann. The last two proved the consistency of 'arithmetics', of that part in which the dangerous axiom about the conversion of predicates into sets is not yet admitted. A gap remained which seemed harmless at the time, but already detailed plans were drawn up for the invasion of analysis. Then came a catastrophe: assuming that consistency is established, K. Gödel showed how to construct arithmetical propositions which are evidently true and yet reducible within the formalism. His method applies to Hilbert's as well as any other not too limited formalism. Of the two fields, the field of formulas obtainable in Hilbert's formalism and the field of real propositions that are evidently true, neither contain the other (provided consistency of the formalism can be made evident). Obviously *completeness* of a formalism in the absolute sense in which Hilbert had envisaged it was now out of the question. When G. Gentzen later closed the gap in the consistency proof for arithmetics, which Gödel's discovery had revealed to be serious indeed, he succeeded in doing so only by substantially lowering Hilbert's standard of evidence. The boundary line of what is intuitively trustworthy once more became vague. As all hands were needed to defend the homeland of arithmetics, the invasion of analysis never came off, to say nothing of general set theory.

This is where the problem now stands; no final solution is in sight. But whatever the future may bring, there is no doubt that Brouwer and Hilbert raised the problem of the foundations of mathematics to a new level. A return to the standpoint of Russell-Whitehead's *Principia Mathematica* is unthinkable. (from: H. Weyl, 'David Hilbert and his Mathematical Work')

The work by Gödel that Weyl refers to in this quote are the Gödel theorems we have been trying to expound in the first chapter of this book.

Chapter III

An Outline of a Complementaristic Approach to Mathematics

> -ὁκόσων λόγους ἤκουσα, οὐδεὶς ἀφικνεῖται ἐσ τοῦτο, ὥστε γιγνώσκειν ὅτι σοφόν ἐστι πάντων κεχωρισμένον- Heraclitus, DK, B frgmnt 108.
>
> [Among all those, whose teachings I have learned, there is nobody who teaches that Wisdom is independent of all common things.]

III.1. FACETS AND METHODS OF A PHILOSOPHY OF MATHEMATICS

In drawing conclusions from the history of the foundations of mathematics, one thing can certainly be said: on account of fundamental (philosophical) differences, the discussions between the different schools of thought have not come to an end.

In fact, if one consults the proceedings of some recent conferences on the philosophy of mathematics (cf. Lakatos) then one can readily distinguish intuitionistic-constructivistic, formalistic and Cantorian points of view.

However, the situation today is not like that of, say, the 1930's, when Gödel's work was only beginning to make itself felt and when the contrast between the schools had arrived at a peak. In continuing where chapter two ended, and incorporating Gödel's results (and others), discussed in chapter one, we can say that there is more unanimity today than there was three decades ago. At least three reasons can be given for this development.

In the first place, the results of Gödel and his school, in particular Cohen's results on the continuum hypothesis, have shown that the traditional axiomatic system for set theory and of disciplines dependent on it are far from complete. That is to say, the results that can be proven or formulated in an axiomatic approach to these disciplines fall far short of what can be formulated in the (spontaneous) natural language of the non-axiomatic versions of these disciplines (cf. Section I.7). Thus the intuitionistic view-point concerning the limited role of language is underscored by the findings of the formalist school.

Brouwer in particular has maintained that the natural, direct mathematical activity cannot be fully formalized and axiomatized. Apparently, Brouwer was dismayed when his disciple, Heyting, formulated a set of axioms for intuitionist logic. However, Heyting maintains Brouwer's mathematical stance, albeit in a slightly modified form. Philosophically, the two men show more divergence (cf. Beth (2), pp. 618-619).

There is, in the second place, a growing awareness that the traditional and ancient dilemmas, such as between apriorism and empiricism, realism and idealism, etc., are misleading and false. However, clarity on these matters has not yet been attained. It is equally true that a return to a moderate realism is taking place, e.g., in the cases of Tarski, Bernays, etc. (cf. Beth (2), p. 612ff).

Thirdly, but tying in with the previous point, it should be mentioned that numerous views have been expressed that lie intermediate between the three general trends of logicism, intuitionism and formalism (cf. Benacerraf and Putnam).

This is in keeping with the *pluralist* view of science which has developed after World War II. Indeed, the pluralist attitude underscores the formation of many interdisciplinary sciences, of which information theory, cybernetics and the computer sciences are but a few examples; and it is typical of this attitude to shy away from any dogmatism that might preclude the development of any branch of science whatsoever. This pluralism, however, does not take away the fact that philosophical problems remain inherent in mathematics, although fewer mathematicians believe in the existence of only one (kind of) mathematics and one (kind of) set theory.

The history of the foundations of mathematics, the present author maintains, can be viewed as an interrelated set of attempts to unite a variety of aspects of the (classical) mathematical activity. Included are the following:

(1) the role of '*logic*', or 'naïve logic', as a spontaneous, innate ability of man to form concepts of things, and an unending variety of concepts of things, concepts of concepts of things, etc.;

(2) the role of *language*, as a communicative tool, bearing information;

(3) the role of (formal) *logic*, as an historically unfolding set of (more or less formalized) rules of thought, made with the specific purpose of ordering and organizing, coherently and deductively, the concepts belonging to specific domains of thought (such as mathematics, physics, etc.);

(4) the role of *constructive thinking*, in the sense of building new entities from given entities in such a manner that the properties of the new entities can be derived from the properties of the old entities, without using a terminology

which (pre)supposes the existence of a 'Platonic' world of *all* properties of some sort;

(5) the role of *intuition*, as an immediate grasping of evidence regarding things, i.e., a way of coming to knowledge, independent of deduction;

(6) the evaluation of the question whether in mathematics there is a 'first knowledge' as opposed to 'derived knowledge', or 'second(ary) knowledge' (e.g., is the number concept primary or is the concept of a 'set' basic to all of mathematics, as Cantorists assume? What is the nature of our knowledge about numbers?);

(7) an account of the fact that along deductive paths the application of purely mathematical methods often leads to (approximate but) true knowledge about the world, i.e., that basic mathematical properties of number and the knowledge of topology and analysis may find application in non-mathematical domains (e.g., is it 'accidental' that the group concept is applicable in physics, or are mathematical structures to be viewed as part of a much wider thought pattern, including thought in the (natural) sciences?).

The author maintains that all these aspects have to be accounted for in a philosophy of mathematics worthy of the name. There are, however, two different ways in which one may deal with these questions, namely, an *implicit* and an *explicit* one.

To *explicate* the function of the above-mentioned factors is to philosophize or *theorize about mathematics*. In the past, most foundational theories have thus evaluated the roles of the different aspects from the point of view of only one, or a few of them, thereby suppressing, disclaiming or neglecting the role of the others. Furthermore, most of the existing foundational theories view all that is usually called mathematics, that is to say, all the mathematical disciplines, even the widely diverging ones, from one central and unifying discipline. When that one discipline or some of its salient features are being considered as typical for the whole mathematical enterprise, then *philosophical reductionism* has taken place.

Here are a few examples. Brouwer's heavy stress on pure mathematics as "constructive thinking proceeding from the basal intuition" made him consign classical (synthetic) geometry from the mathematical domain to that of the experimental physical sciences, thus leaving only the discrete point of view for mathematics as a whole. Another example, which Bernays has repeatedly drawn attention to (cf. his discussion of Wittgenstein's work in Benacerraf and Putnam), is that the formalist trend in mathematics and in the foundational theories has led to a neglect of (the philosophy of) geometry. He remarks, in the work referred to, that contrary to linguistically inclined philosophers of mathematics (such as Wittgenstein), it remains a fact that the classical theorem "the sum of the three angles of a triangle is equal to 180°" can be 'experimentally' verified by an increasingly refined system of measurements. In Lakatos (1), Bernays mentions that any sum, exercise or calculation may be regarded as an 'experiment' in abstract analysis or algebra, even though the meaning of the word 'experiment' has to be redefined somewhat to distinguish it from physical experiments.

More generally, in a sense, every theory has a tendency to simplify. It is unavoidable that where the complication of a domain of interest is greater, the simplification is also bound to increase. Thus certain features are accentuated and others deleted. It is understandable why the main trends of formalism and logicism select the formal logical aspects of mathematics as their vantage point rather than an experimental-intuitive-constructive one. However, in trying to account for this tendency, we meet again the basic problem mentioned in the discussion of Descartes' ideas, namely, the distinction between synthetic and analytic methods, between construction and deduction.

The *implicit way* of dealing with the basic aspects of mathematics begins with the *working mathematician*. A mathematician working in domains such as number theory, topology, algebraic geometry, etc., can never avoid any of the seven aspects mentioned above. They are all implied in what he does, in the sense that he deals spontaneously with these moments, especially when the worker is busy in disciplines which are a direct extension of the classical mathematical disciplines. For instance, it is unthinkable that anyone working on problems in (algebraic) number theory would not use his 'intuition' of smaller-and-greater-magnitudes. Accordingly, the mathematical reasoning he 'applies' is closer to what we called 'logic' in the sense of (1) than formal logic in the sense of (3). Naturally, as the historical development of modern mathematics shows, these different ways of applying logic may not be completely separated. But our concern is that, even though workers in the extended classical fields have to choose from time to time which (weak or strong) set theory to use, there is still a particular nature to their activity which is all too easily overlooked when one simplifies the multitude of stubborn mathematical disciplines into a narrow (formalist or intuitionist etc.) philosophy. Creative, 'naive' and spontaneous mathematical thought has to be sharply distinguished from the formal logical and foundational reconstruction of that which results from spontaneous thinking.

III.2. TWO KINDS OF MATHEMATICAL EXISTENCE

We would now like to express our opinion as to how we think the above-mentioned aspects hang together in mathematics. As a first approximation, we say that the results should be different for different disciplines.

The following examples will show this clearly. It has already been remarked that giving a consistent priority to the constructive-intuitionistic aspects of mathematics (Brouwer) necessarily leads to the abandonment of synthetic geometry and the classical concept of the geometrical continuum. (Formalism and logicism never meant to abolish the classical continuum concept, nor Cantorism, at that; however, the latter did try to derive the geometrical continuum concept from the discrete set concept.) In view of this, is it not justified to say that Brouwer's idea of constructivity applies only to those disciplines which start from the discrete, such as number theory, algebra, numerical analysis, etc.? It is well-known that Brouwer rejected any reference to a 'complete' continuum as a usable concept, calling it 'theological' or 'metaphysical'.

By the same token, is not Bernays' observation on the lack of interest in geometry by logicians and philosophers indicative of the fact that there is not just one kind of existence in mathematics, and that to view all of mathematics in the light of one concept of existence denatures those disciplines which depend on others? If one answers this question in the affirmative, then one may appreciate Brouwer's criticism of classical analysis and geometry in this sense, that one admits that in mathematics *two different kinds of existence* occur intermingled, but legitimately, namely, a 'geometric' and a 'discrete' one. Taking this view, one is compelled to answer the following questions:

(1) Does there correspond, to the two different kinds of mathematical existence, two different set theories? It is clear that this question ties in with the so-called doctrine of the 'arithmetization of geometry', which says that to every classical geometrical statement, there corresponds an arithmetical one. One notes immediately that this doctrine is a corollary of the Cantorist's ideal to rebuild continuous concepts from discrete ones, or geometric concepts from that of a set.

(2) If there are two different kinds of set theories at work in classical geometry and analysis, then do these set theories depend on the 'nature' of the entities which are being collected into a whole? How can one distinguish these set concepts?

Before giving a sketch of how we could conceivably answer these questions, we would like to mention an additional motive that determined our choice of view on foundational matters in mathematics. Reading through the articles of H. Weyl, especially those on Hilbert's formalism (cf. Weyl), it is telling that Weyl could not in the final analysis agree with Hilbert, precisely because the latter believed that beyond the domain of the (intuition of the) discrete, there is a domain of continuity, which the mathematical mind is equally able to grasp and to operate with (Weyl, Vol. 4, p. 330ff., p. 599ff.). (It is well-known that H. Weyl's growing disbelief in such a domain made him support Brouwer's views on constructivity. It is equally well-known that Brouwer termed the physical world of change and continuous motion vague, because it lacks the clear exactitude of the intuition of two-oneness.) This view of Hilbert seems all the more important because for Hilbert the pure concept of continuity provided a connecting link between pure mathematics on the one hand, and applied mathematics and physics on the other. Indeed, according to Brouwer and Weyl, to speak of a pure concept of a (linear) continuum, is to introduce into mathematics the *actual infinite*. They conceive of the continuum as something potentially constructable from out of discrete notions (as medium of free choices). In fact, this concept of a continuum concurs with the Cantorist one, in that the latter also conceived the continuum as constructed from discrete notions, but differs from it in that Cantor accepted the hypothesis that the human mind is capable of constructing, beyond the 'potential infinite', the actual infinite.

Now, motivated by our wish to link the 'applied' and classical geometrical aspects of mathematics to pure mathematics, we surmise that one may well speak of the concept of a geometric continuum without postulating that the human mind is able to construct actual infinities. To do this, we have only to resort to the view that the concepts of point, line (-stretch), plane, etc., can be directly conceived as 'geometric entities', and that their existence is *sui generis*, i.e., independent of any constructive ability which puts lines together from points and the like. Thus, we do not say that entities of the geometric sort can be 'picked up' from the physical world like (the concept of) physical things; on the contrary, it seems that these concepts are at least as much 'derived from the world' as the elementary arithmetical ones are. But, nevertheless, we do say that we can pick them up! The assumption that there are (at least) two distinct fundamental ways of obtaining the basal concepts of discreteness and continuum is fully supported by epistemological investigations such as those of Jean Piaget (cf. Piaget). These investigations confirm that the so-called space intuitions develop, in a manner relatively independent of the natural number concept, during the maturation process. In fact, Piaget calls a line, a curve, etc., 'spatial symbols', standing for actual bodily physico-spatial transformations. If we understand Piaget well enough, then he thinks of the purely spatial conceptions of the human mind as being the result of physical motions and transformations. As it is here not the place to try to follow Piaget's 'épistémologie génétique' in detail, we limit ourselves to express that Piaget's findings alone legitimize the hypothesis that mathematics is based on at least two basal intuitions, the intuitions of discreteness and the intuition of continuity, and that each corresponds to a particular kind of existence, namely, existence in the sense of number and discreteness, and existence in the sense of continuum and spatiality.

We are now in a position to state the complementarist position with respect to the existence of the basic 'entities' number and continuum. Our position is that they are *qualities* of the world we live in. This has to be explained further, because we obviously do not mean to say that they are sensory qualities of one kind or another. And neither do we say that they exist in a Platonic world of perfect entities, because they are *not* entities, objects, substances or 'things'. There is, for all practical purposes, no objection to mathematical theories handling them, or speaking about them, as 'entities', even though their genetic origin does not warrant this terminology. Thus, the number '5' is certainly a quality of, for example, the set $\{a, b, c, d, e\}$. Further, if line segments etc. were not qualities of things, then Bernays' remarks on 'experiments' regarding triangles (cf. Section III.1.) would not make any sense at all. Put in other words, the *mode of being* of numbers and lines etc. is that of a quality. We reserve the term *analytic quality* for them in order to denote that picking them up from the world presupposes analytical activity on the part of the human mind, though any epistemology accounting for their learning process seems to only touch upon what really happens. Their mode of being, being that of qualities, thus excludes (Platonic) questions such as: where do they exist, are they perishable, etc.? One might say that as long as there is matter, there is continuity and discreteness, even if there is not anyone around to have a concept about them. Their concept, in turn, is not a priori, but a posteriori, while the ability of the human mind to grasp these qualities is *innate* (and differs from person to person).

But what is to be said of, for instance, the existence of lines, curves, etc. in the context of a *geometrical theory* (which handles them as 'entities')?

First of all, it is, for the purpose of building a *purely* mathematical theory, entirely legitimate to speak of these 'entities' as if they are objects, as it is not the purpose of a pure mathematical theory to keep these objects, as qualities, constantly linked to the world. It is rather *applied* mathematicians who have a tendency to try to keep track of the physical meaning of the parameters in their formulas, thus handling numbers and geometrical entities as qualities. The history of mathematics, as a history of the emancipation of pure mathematics from the applied sciences, is marked by the struggles of geometers to justify the 'geometries' they defined, either in terms of physical application, or in terms of newly constructed 'models' made with geometrical entities as some sort of building material.

Secondly, it seems entirely wrong to restrict the concept 'geometrical entity' to three-dimensional ones, because even physics provides us with plenty of examples of magnitudes that are dependent on more than three variables, thus making it possible to conceive of higher dimensional qualities. This does not take away, however, the fact that in passing from

3-dimensional geometry to n-dimensional geometry ($n > 3$), the human mind experiences that in lower dimensions he can make use of representations by means of his sight senses, whereas in higher dimensions he can do so less.

In the third place, in view of the history of empiricism, it seems necessary to emphasize the fact that the human mind has an innate ability to mentally manipulate and freely operate with the basic geometrical entities. He does not generally handle them as if they are autonomous sources of knowledge, but he seems to prefer to build geometrical 'models' from them, in which the axioms of particular geometries are satisfied; he thus passes to new kinds of experience and experiments in a world that he himself has helped to create, relatively independent of the physical world.

[Here we meet a situation which is common to all pure mathematical disciplines. An axiom system, defining a particular mathematical 'structure' (a non-Euclidean geometry, a sheaf, a manifold, rings, groups, etc.) admits in general more than one constructive 'model', even though, genetically, the axioms were formulated with a view to one or only a few important examples of models, that may or may not have a link with physics. It is typical of pure mathematics that it is interested not so much in the individual entities from which models are constructed, as in the models as a whole. Thus if in ring theory, individual numbers, or discrete entities constructed from them (polynomials, functions, cosets, etc.), are the object of study, then the interest in them usually, leads to, stems from, or serves the study of the more general properties of the models that are made up of these numbers and entities. This feature is one which distinguishes mathematics from the natural sciences. The physicist is very much interested in the structure of the atom (and hence, of every atom). We cannot say that in a similar fashion the mathematician is interested in the structure of a line segment, number, polynomial or function. On the contrary, the structures mathematicians are interested in are *sets* of lines, numbers, etc., the entities which form the set being interpreted by the rules holding for the set (field, rings, particular geometries, etc.). On the face of it, though, it does not seem to be true that the two principal physical disciplines are dealing with the structure of atoms etc., namely, quantum mechanics and the theory of relativity. The historical development of both of these subjects, however, shows that the increase of their abstract mathematical methods was rather necessitated by the limitations of physical experimentations and measurements, than by the belief that there is no structure to atoms, molecules, etc. In the former case, physicists were forced (the Heisenberg relation) to resort to probabilistic methods, which enabled them, in some cases, to convert information on multitudes of particles to information on the structure of the particle. In the theory of relativity, which is fundamental for Einstein's effort to come to a unified field theory, the underlying idea has been the philosophy that all physical events can be reduced to and described in terms of (physical) space and 'physical geometry' (cf. Wheeler), thus in terms of the continuous: *particles are being conceived as particular discrete topological configurations of physical space satisfying certain differential equations of geometro-dynamics*. It is recalled here that the decades-old dispute between the proponents of quantum mechanics and of the theory of relativity has only been 'resolved' by Bohr's *principle of complementarity*, whereby one views continuity and discreteness as two different aspects of one and the same physical world, competing to form the vantage point of physical theory.]

As a final point in this paragraph, let us touch on the status of the real numbers. For, up till now, we have only been dealing with the positive integers, or the natural numbers, as qualities. To see that the real numbers can be viewed as 'qualities of qualities', and hence, as having the mode of being of qualities, fix a unit interval on any straight line. Then the Lebesque measure defines this line to become the usual real number axis. In the light of what has been said before, the number '5' is then a quality of a well-defined point (or of a well-defined line-stretch). As, even from our non-Cantorist viewpoint, one may say that "the line contains the set of its points (cf. Section III.3), the former statement on the number '5' generalizes to saying that any arbitrary real number (as a convergent series of rational numbers, for example; or as a decimal fraction) is a quality of a well-defined point on our number axis. One could speak here of a 'qualification of qualities'. A similar reasoning applies to complex numbers and to Euclidean n-space, where n-tuples of real numbers can be viewed as qualities of points.

Before trying to answer the two questions posed in the beginning of this paragraph, we have to discuss in more detail the relation between the continuous and the discrete.

III.3. LANGUAGE, SET THEORY AND MATHEMATICAL COMPLEMENTARITY

Before discussing in more detail the relation between the two different basic concepts of mathematics, first a few words on the role of language. We have indicated that formulating a system of axioms for a particular discipline means creating a language that is relatively loose from the domain of entities ('models') to which that language may apply. In fact, this should not be taken to mean that we consider the models to be pre-lingual, in the sense that mathematical thought about them, and the language of the axiomatic theory which brings out that thought, *follows* the construction of the models, in a fashion which is sometimes called 'the natural function of thought' (cf. Beth (2), last chapter). It occurs, on the contrary, that the actual interplay between the three mentioned functions 'construction', 'not yet formulated insight or thought' and 'axiomatic language' is such that at times each one of them precedes the others. An outstanding example of formal axiomatic reasoning preceding the construction of models is, of course, provided by the history of non-Euclidean geometry. But in a slightly different fashion every system of axioms defining a general structure invites the question in which model those axioms are satisfied. We speak of an *anticipatory function of language* in this connection. This 'running ahead' of language is also evident when the manipulation of a formula (or a set of formulas) leads to a result that could not be thought to be true before the manipulation. This may occur in every level of mathematical thought. It has been Brouwer who has made the severest objections against this role of language, even to the extent that he required that language should only *follow* the (pre-lingual) constructions, and that language should be stopped once it no longer follows closely the track of those constructive operations. This brings us to consider the point, against which the thrust of Brouwer's critique was directed, namely, his consideration that the existence statements of (classical) analysis, insofar as they were not 'covered' by constructive-arithmetical statements, were of geometric origin, in the classical sense. Theorems such as "a continuous function on a closed interval attains its maximum value", the theorem of Bolzano-Weierstrass, etc., make use of quantifiers running through infinite sets of intervals and points, irrespective of whether or not these sets can be constructed. They are examples of theorems that had to be modified in neo-intuitionism. A more crude example is obtained when one considers "the set F of all continuous functions on the unit interval (with values, say, in the unit interval)". There is no constructive arithmetical device that will produce 'all' the functions with this property, if one understands by 'arithmetical device' a way of actually writing down a 'numerical law' which (supposedly) will generate the individual numerical expressions for 'all' the functions on the unit interval that the classical-geometrical intuition may imagine. Moreover, there is no geometrical device known that will be capable of generating all these functions, as well. Whence the question: when may we say that a certain set of (geometrical) objects has been formed, so that one may conceive it as a new mathematical entity?

[Restricting our attention for a moment to sets of geometrical objects we briefly take the opportunity to explain the difference between constructing continua from out of the discrete on the one hand, and, on the other hand, by conceiving them directly via kinematic geometry and physics (and, perhaps later attach to them discrete algebraic notions). As regards the latter constructions, there exists no complete catalogue of functions that can be obtained as compositions of motions of a point, say in the plane, but here are a few examples: straight lines, the circle, the sine, the exponential function and branches of polynomial functions. (The reader may convince himself that the perpendicular composition of two motions, one of which is uniform and the other suitably uniformly accelerated, gives rise to a branch of functions of the form $y = x^n$, etc.). Further, as regards the former constructions, i.e., the constructions of continua from out of the discrete, if for a real function f, the formula $y = f(x)$ is explicitly given, then it can be plotted in the plane and it is said to define a *plane function* (possibly under conditions that restrict the range of the variable x).]

Now, our main point as far as the formation of sets is concerned, is the epistemological requirement that the *shape* of the objects that are going to form a set has to be made clear *in advance*. The word 'shape' then, may be taken in the geometrical-kinematical sense as well as in the arithmetical sense. In the former case, for instance, the plane 'figures' point, straight lines, circles, the sine, the polynomial etc., have a well-defined shape, obtained from kinematic geometry. In the same fashion, in the latter case, does one define plane functions from well-defined expressions in calculus, using number symbols, variables, and various properties of (real) numbers and it is clear that one can define more plane functions via calculus than from kinematics. (Incidentally, this may be one of the reasons why geometry in the classical sense has been pushed to the background, cf. Bernays' remark in Section III.1.) In both cases may one only collect into a set plane functions that are special cases (or specialisations) of general shape-constructions in (kinematic) geometry or in calculus, respectively.

An example of a general shape-construction in geometry is the circle (viz., as the locus of all points in the plane having a fixed distance to a fixed given point). Furthermore, the sine-curve can be viewed as the composition of two motions, one of which is uniform along a straight line and the other one uniform on a circle. These general shape-constructions contain variables (a fixed, but unspecified distance, a fixed but unspecified point, etc.) that can be specialized and once this is done one obtains special constructs. In view of what has been said before it is, epistemologically, thus permitted to speak of the set of *all* such constructs, and hence of the set of *all* circles in the plane, the set of *all* sine-shaped curves in the plane etc., whereas one cannot speak of the set F defined before. Likewise, an example of a general shape-construction in arithmetic or calculus is the polynomial function, say with integer coefficients and of degree n. Its general shape could, for example, be defined by the expression

(*) $\quad a_0 + a_1x + \ldots + a_nx^n,$

with a_i $(0 \leqslant i \leqslant n)$ real number variables and x the polynomial variable. Thus, the integral numbers forming a set, one may speak of the set of all polynomials of degree n with integral coefficients. The reader easily adds numerous examples to this one. He should also note that the notion of a wff (well formed formula, cf. Section I.5) of the Predicate Calculus is an example of a 'shape-construction', but then in Formal Logic. One may thus speak of the set of all wff's and check through an infinite list of wff's as in the proof of Gödel's theorems (cf. Section I.7).

If we now compare the complementarist's way of forming sets with the Cantoristic(-realistic) and constructivistic (-idealistic) ways, then the first thing to remark is that in Cantorism totalities such as 'all plane functions' etc. may be called *sets* because one has available the extremely powerful axiom of the 'power set' (saying that if V is a set, then the collection of all subsets of V is a set). Such an axiom does not follow in any sense either from the epistemology of constructing plane functions from out of the continuous, or, indirectly, from out of the discrete. On the other hand, requiring that for the conception of a set it is necessary to first give a general shape-construction is not at all constructivistic, just because there is still a difference between 'existence' and 'constructibility', as has been explained in the above. Thus, because there is still an epistemologically unfillable gap between the discrete and the continuous, it is a characteristic of complementarism to regard only that part of mathematics as *epistemologically valid* that respects the relative independence of the discrete and the continuous throughout the natural sciences. (Any axiomatic theory that uses tools more powerful than those implicit in what is said above, may, *mutatis mutandis*, be called *deductively valid*, on the condition that it does not contain contradictions.) That is to say, it is typical of complementarist thinking to regard the discrete and the continuous as *complementary aspects* of the world, in a fashion quite analogous to the complementarity principle in physics: depending on the type of problem that is being attacked in mathematics, we either tend to assume existence after the geometrical-continuous fashion, or we tend to require existence or computability in the sense of numerical-discreteness. And this is exemplified by the fact that some mathematical disciplines depend heavily on the (intuitive) 'handling' of geometrical continua (Euclidean n-dimensional topology, classical geometries, etc.), whereas, at the other extreme, some disciplines only deal with (sets of) discrete entities (algebraic number theory, numerical analysis, etc.) (cf. also Section III.6). In Section III.6 we shall divide the principal mathematical subdisciplines into two categories, viz. the category of those that have their (epistemological) origin and final purpose in the domain of the discrete and number theory, and those that stem from, and are intended for, the domain of the continuous and geometry.

The main question here to be settled now is the role of set theory. As set theory is not a theory which reduces the continuous to the discrete, we have not much choice left than to regard it as a theory formulating rules for handling geometrical as well as arithmetical entities. Hence, on this view, *set theory regulates the usage of the mathematical language as it bears on the formation of new totalities of entities from given totalities of entities*. Thus, set theory obtains a normative character pertaining to the mathematical language.

The fact that set theory formulates a *norm* for the usage of language is already evident from the fact that set theory is the only mathematical discipline in which, historically, problems have arisen of the type: "*to what extent* can one assume the existence of mathematical entities?" Basically, the answers given to this question went in two different directions: (a) Cantorist-realist-formalist attitude, which maintains that one could go as far as one likes, on the condition that no contradiction results in the system; and (b) the remaining attitudes, including the idealist-intuitionistic ones and the logicistic-realistic ones, which delimit the language, and let set theory depend on cosmólogical considerations (cf. Sections II.4.1 and II.4.2).

From the above, it is obvious that, if we were to formulate more explicitly a set theory which agrees with the complementarity principle, then we are to be counted as belonging to the second class of attitudes (b).

Furthermore, our wish to adapt a viewpoint which balances the seven mentioned aspects of the mathematical activity in such a fashion that none are suppressed in favour of the others (Section III.1), makes us believe that set theory must make explicit the implicit 'naive set theory' operative at the (epistemological) level of not-yet-fully-formalized mathematical experience. This means, once more, that we support the *a posteriori* view of Piaget, according to which during the course of the mathematical learning process, the individual human mind passes through a set of natural levels of abstraction, in such a fashion, that every 'later' level of abstraction is dependent on the 'former' ones (more or less analogously to the historical development of mathematics). Indeed, if we were to state what makes mathematics such a 'sure' and indubitable science, then we would have to point out that its indubitableness does not simply flow from the abstract, formal or axiomatic, method, but that even the abstractions are made possible by, and owe their existence to the relatively simple and clear (as compared to the concepts of the natural sciences) basic mathematical intuitions of continuity and discreteness. The developments as described in Section I.8 have done nothing to support the original logistic and formalistic trust in the axiomatic and formal methods, even though these methods are very appropriate for the formulation of purely mathematical theories.

III.4. COMPLEMENTARIST SET THEORY–AN OUTLINE

The complementaristic attitude is given to the desire to respect the naive, spontaneous, set theory of the epistemological level and of classical mathematics and, *ipso facto*, to respecting the original unity of pure and applied science. In the previous paragraph it has been explained that inherent in the classical mathematical attitude is the trait that the definition of sets of entities *follows* a general shape-construction rather than that sets are defined just by means of giving a property, and then take such a set to be a member of a further set, etc. Historically, only when mathematics became more abstract and the tendency grew to build sets of sets of sets etc. of sets of, say natural numbers, did it become evident that those who opted for an a priori view of mathematics were compelled to make a choice between a (philosophical) attitude that either tended in a Cantorist-realist direction or in a constructivistic-idealist direction (cf. Section

II.3). In terms of the quotation from Poincaré (cf. Section II.4.2.1): it appeared as if the a priori view compelled mathematician-philosophers, at least in the matter of existence questions in mathematics, to either direct their search-light to the world of the continuum and geometry (Cantorism, etc.) or to the world of the discrete and number theory (constructivism, etc.). The complementaristically minded mathematician, who considers mathematics, a posteriori, to be built up from 'first material' (natural numbers and the continuum as analytic qualities of the world), and who holds the further view of set theory as formulating a language *norm*, is not placed in such an either-or-position. Because neither does the term 'built up from first material' mean equating existence and constructibility (as in constructivism, cf. Section II.3), nor do 'general shape constructions' give rise to epistemological outlandish predicates or properties (as in realism-Cantorism).

Let us, before discussing the possibility of an axiomatisation, first remark that, in passing, we have already given an answer to the two questions posed in Section III.2. Indeed, in the course of Section III.3, when describing the way in which sets were built in the realm of the continuous (via kinematic geometry) or in the realm of the discrete (via formal expressions), we did discuss the general epistemological pattern common to *both* realms. Whence a negative answer to both questions of Section III.2: the complementarist view does *not* lead to two different kinds of set theory, one pertaining to the discrete and one to the continuous. Thus, the next question becomes: is it possible to formulate the complementarist position in an *axiomatic fashion*? It is clear that, on account of the totally different nature of set theory from the other mathematical disciplines, one may not expect a clear-cut axiomatic theory that fully takes care of complementarism. Incidentally, the clear-cut Cantorist axiomatisations of set theory (e.g. the Zermelo-Fraenkel system ZF and the Gödel-Bernays system GB), were, historically, designed to fill up the epistemological gap looming between the discrete and the continuous, and these theories may well be deductively valid (cf. Section III.3), even though they are not provably so, on account of the Gödel results; it is constructivism and especially neo-intuitionism that started to cast a serious doubt on the possibility of a complete and adequate axiomatisation of any mathematical theory and of set theory in particular, to begin with. These circumstances, however, cannot prevent us from trying to indicate an axiom system that reflects as much as possible the main traits of the complementarist view, as pictured in the previous paragraphs.

Before summing up the axioms that could serve as such, let us recall that, generally, in order to define a set one needs a general shape construction (cf. Section III.3). The question becomes whether for instance the real numbers form a set. The answer is yes, because, first of all, the (positive) integers do form a set since they constitute, as arithmetical primitives, the 'first material' in arithmetics. Then, as a general shape-construction for the reals one may take $n, a_1 a_2 a_3, \ldots,$ where n is an integer variable and a_1, a_2, a_3 are (decimal) variables running through the set $\{0, 1, 2, \ldots, 9\}$. There are, of course, other ways of giving shape-constructions for the reals: we may view them also as convergent series of rationals, etc. In a similar fashion one may speak of the set of all straight lines in the plane, the set of all points in 3-space, etc. simply because we know what a straight line, point, etc. is: a primitive geometric concept that can neither be reduced to something else nor explained in other terms.

In view of all this, it will come as no surprise that the main point at which complementarism deviates from the Zermelo-

Fraenkel (ZF) system is the axiom of the power set. The construction of the real numbers out of the integers and rationals hinted at previously, suggests the replacement of the axiom of the power set by the following one:

A_7 *Axiom of Specialisation*

Let X be any non-empty set, then there exists a set $P_0(X)$ consisting of all countable (i.e., finite and denumerable) subsets of X.

The fact that this axiom reflects the construction of the real numbers is better seen if we formulate it in a form which is equivalent to it, within ZF:

Let $X = \{x_0, x_1, x_2, \ldots\}$ be a denumerable unordered set of sets with $x_i \neq x_j$ for $i \neq j$. Let f: $X \to Y$ be a bijection of sets. Then there exists a set $P_0(X, Y)$ consisting of all (unordered) sets of the form $\{z_0, z_1, z_2, \ldots\}$ with $z_i \in f(x_i)$ for all $i \in \mathbb{N}$.

The role of the sets x_i is analogous to the role played by the decimal variables before. We list the further axioms, not in fully formalised (within the predicate calculus) form because it would not serve our purpose here. For a full discussion of the formalised axioms the reader is referred to Cohen (Chapter II).

A_1 *Axiom of Extensionality;*

saying that a set is determined by its members:

$\forall\ x, y\ (\forall z(z \in x \leftrightarrow z \in y) \to x = y)$.

A_2 *Axiom of the Null Set (*= empty set*);*

the null set ϕ is defined by $\exists x\ \forall y(\sim y \in x)$.

A_3 *Axiom of Unordered Pairs;*

Given x, y sets, the unordered pair is denoted by $\{x, y\}$ and defined by the formula

$\forall x, y\ \exists z\ \forall w(w \in z \leftrightarrow w = x \vee w = y)$.

A_4 *Axiom of the Sum Set or Union;*

Given a set x (the members of which are sets) then one can take the union of all these members on account of:

$\forall x\ \exists\ y\ \forall z(z \in y \leftrightarrow \exists\ t(z \in t\ \&\ t \in x))$.

A_5 *Axiom of Infinity;*

Let x be a set then the 'successor' of x will be defined by $x \cup \{x\}$, as follows:

$\exists x(\phi \in x\ \&\ \forall y(y \in x \leftrightarrow y \cup \{y\} \in x))$.

Actually, this axiom guarantees the existence of the integers in ZF, because in ZF the integers are *identified* with the set

$\phi, \phi \cup \{\phi\}, \phi \cup \{\phi\} \cup \{\phi \cup \{\phi\}\}$, etc.

Thus the nature of the positive integers as qualities is completely lost, but we do not see how to salvage that nature in *any* axiomatic version of set theory.

A_6 *Axiom of Replacement;*

We do not give a within ZF formalised version of

this axiom. Its content concurs completely with the idea of general shape-constructions. The axiom says that, given a formula $A(x, y; t_1, \ldots, t_k)$ in predicate calculus which contains at least two free variables, and which defines y uniquely as a function of x ($y = \phi(x)$, say), then for each set u the range of ϕ on u is also a set. The general shape-construction is here the function $\phi(x)$, which is implicitly given by the predicate A.

A_8 *Axiom of Regularity*;

This axiom states that each non-empty set x contains an element that is minimal with respect to the relation ϵ, and may be formalised as follows:

$\forall x \exists y(x = \phi \vee (y \in x \,\&\, \forall z(z \in x \rightarrow \sim z \in y)))$.

A_9 *Axiom of Choice*

Let X be a countable set, $a \in X$. If $a \leftrightarrow A_a \neq \phi$ is a function defined for all a, with A_a sets, then there exists another function $f(a)$, and $f(a) \in A_a$.

This is not the usual axiom of choice occurring in Cantorist set theory, because of the condition that X be countable. The present axiom is in line with the approach of Section III.4., 5 because, in fact, the function f defines a vector $(f(a))_{a \in X}$ for which one can give a general shape construction (e.g. $(x_a)_{a \in \mathbb{N}}$), with x_a variables), whereas it cannot be seen that such a construction is possible for infinite sets X that are not countable: in the level of the construction of symbols on paper it seems that one can only construct countably many distinguishable symbols or variables.

Whenever an axiom system for set theory is being given,

the main question becomes: to what extent can the extant mathematical disciplines be formulated with the help of that set theory? As far as the set theory based on A_1-A_9 is concerned, not too much is known. That is to say, it is common knowledge that one can derive with it what is known under the name 'classical analysis', but there remain questions, hitherto unsolved, as to what parts of, for instance, the general theory of Analytic and Differential Varieties can be retained. The difficulty lies in the fact that one does not yet have a sufficiently good idea of which functions (on $\mathbf{R}$, say) can be defined by means of A_1-A_9 inside the hierarchy $P_0(\mathrm{N})$, $P_0P_0(\mathrm{N})$, $P_0P_0P_0(\mathrm{N})$, etc. In the same vein, on the algebraical side, it is not at all clear what parts of 'abstract' Algebraic Geometry and the 'abstract' theory of Banach Algebra's can be retained, mainly because from A_1-A_9 one cannot draw a convenient weakening of Zorn's lemma (which, in ring theory, is to the effect that every ideal is contained in at least one maximal ideal). Note that with A_1-A_9 not every inductively ordered set of subsets of a set need have a maximal element, so that, here too, one seems to be bound to sacrifice part of the general theory by restricting its validity to schemes and function algebra's that can be constructed by way of A_1-A_9. In both the analytic and the algebraic case, however, no techniques that are essential to the diverse disciplines are being affected by this 'weaker' set theory. Before discussing the division of the mathematical disciplines into algebraical and analytical-topological ones, first a few words on the unity of mathematics.

III.5. THE UNITY OF MATHEMATICS: ALGEBRA AND TOPOLOGY

The set theory resulting from A_1-A_9 formulates an epistemological norm in this sense that it makes it possible to get a view of the unity of that part of (pure and applied) mathematics that is of epistemological import: as soon as the axioms A_7 and A_9 are being replaced by stronger ones – such as for example the unrestricted axiom of the power set and the (unrestricted) axiom of choice – then, in order to conceive the unity of the mathematics built from them, one has hardly any other choice than to resort to the view that mathematics is but a deductive science, a priori to all experience.

Of course, constructivists would tend to agree with the latter conclusion, but they would try to save the a priori position of mathematics by stating that, since the mathematical entities are constructions of the human mind, there is no epistemology at the basis of mathematics at all and, hence, that one should refrain from believing in an intrinsic connection between the pure and applied aspects of mathematics; for the order in nature is rather a result brought about by the ordering human mind, than *vice versa*.

Complementarism walks the tight-rope between an unrestricted Cantorism-realism and a narrow constructivism, without being given to the desire to force all mathematics into a strait-jacket axiomatic system. Thus, it believes that there is not just one level of mathematical exactness, but that there are many ways and many levels in which particular branches of mathematics may be perpetrated, just because the basic mathematical entities (or better: qualities, cf. Section III.3) and their composition rules, available to almost everyone, are not in need of being clarified or justified by any theory, while there is no existence problem about them. It believes that a constitutive part of the pure mathematical activity is: proceeding constructively with these basic qualities as 'building material', and it believes that the resulting constructions (functions, spaces, sheaves, etc.), for the most part being the product of human activity, have a mode of being that is *derived* from the mode of being of the basic qualities. In terms of the set theory this means that, on this view, the derived entities may not be allotted an existential status that exceeds the status of the basic qualities of the real numbers; whence the approach to set theory taken in Section III.4. Thus, in the level of the existence of mathematical entities one could say that the mathematical constructions lead a *derived existence* as opposed to the *original existence* led by the basic qualities.

Complementarism further believes that the second constitutive part of the mathematical activity is the formulation of a specific form of *knowledge*; and it believes that that knowledge is knowledge *about* the mathematical constructions (including the basic qualities), as well as knowledge about the (physical, biological, etc.) world, viz. in so far as mathematical results, via the basic qualities, can be connected (or: reconnected) with the (physical, etc.) world.

The formulation of that knowledge then takes place in deductive patterns, and the knowledge that results is mostly indirect. As an example take Ring Theory. From the axioms for a ring and with the help of suitably chosen definitions one may deductively derive ring theoretical knowledge. Here, as well as in general, the axiom system and the definitions that one employs, are given with a view to one or a few particular *examples* or *special cases* (of rings, etc.) that one has in mind. And it is not uncommon that one keeps that special case in mind throughout the deductive process in a fashion analogous to what Descartes viewed as the role of the intuition (cf. Section II.2.3.2): one proceeds in the intuition of the special case. The result of the deductive process is more often than not knowledge valid for a very great variety of cases, including the original one, so that one may say that one has arrived at knowledge that is more general than the knowledge one had before. Thus, the playing around with the particular examples or special cases one had in mind, can, in a very true sense, properly be described as *mathematical experimentations*, and it might have been also this sort of experiments that P. Bernays had in mind when speaking of mathematical exercises as experiments (cf. Sections III.1, 2). The history of mathematics of the last hundred years, at any rate, is an historically unfolding process of deductive processes of this kind, in which the two constitutive parts of the mathematical activity play a complementary part: the particular cases or models the deductively formulated theory is dealing with, are constructions wrought with set theory and it is the particular concern of complementarism to point out that those theories that only admit models that can be formulated by means of stronger set theories than made with A_1 - A_9, have less epistèmological import than those theories that admit models constructed by means of A_1-A_9 as well. Non standard analysis (Section I.8) is a case in point.

So, if the complementarist does not conceive the unity of mathematics as 'constructions from out of the discrete' (like in constructivism) or as 'deducibility from systems of axioms', then what, for him, makes mathematics the unity it appears to be? The answer is that mathematics is the exclusive science that studies the interplay and the connections between the mutually irreducible qualities of the discrete and the continuous, of number and spatiality. This interplay and these connections are not furnished by axiomatic decrees such as the axiom of the power set and the axiom of choice, but it is the *mathematical disciplines themselves* that by their methods and results establish these connections. To illustrate this one has to call to mind the incisive difference inside mathematics between algebraic methods and methods of an analytical-topological nature.

By this token, it is not an accident that *Algebraic* Geometry is that branch of Geometry that is most suitable for applications in Algebraic *Number Theory*. In the same vein, it has been Geometry of Numbers which, by laying the connection between lattices (discrete) and analytic functions (continuous), provided the historic link between Algebraic Number Theory and Analytic Number Theory. It is further not merely a fluke that there is Combinatorial Topology (discrete approach) and Analytical Topology (continuous approach), Combinatorial Probability Theory and Analytic Theory of Probability (cf. also Freudenthal). Many examples may be added to these but here we recall rather from the recent literature the famous work by P. Deligne, who recently (1973) proved the Weil-conjectures (cf. Serre, Dieudonné). If there is any excellent example of mathematics evidencing the interplay and exchange of methods from disciplines belonging to the domain of the discrete and the domain of the continuous – with the ultimate goal to obtain arithmetical information – then here is one. The topological methods that play an essential role in the proof of the conjectures, are adaptations to the number theoretical domain, of the so-called Zariski-topology, which itself is a topology appropriate to the domain of Commutative Algebra. It is the so-called *étale topology* on the sheaf defined by a non-singular projective hyper-surface defined over a finite field. The Weil-conjectures then, concern the ζ-function of such varieties, asking for the number of solutions of a set of defining equations of such varieties, these numbers themselves figuring as numbers of fixed points of the powers of the Frobenius-automorphism acting on the (algebraic) points of the hyper-surface.

In a more general fashion, the counterpointal make-up of the higher theories does support the view that the standard mathematical disciplines, i.e., those that grew out of the classical disciplines, furnish *themselves* the tools that build bridges spanning the abyss between the discrete and the continuous. Complementarism is therefore inclined to say that the 'duality', discrete-continuum, algebraic-topological, number-spatiality, etc. is a *structural* one, given the world as it presents itself to the human mind. The set theory defined by A_1-A_9 can be viewed as a tentative attempt to give a formalised expression of where the borderline lies between the epistemologically valid part of mathematics and the deductively valid part (cf. Section III.3).

We finish this section with two remarks intended to clarify the complementarist-philosophical position and its language further as it relates to the history of mathematics. The first remark concerns the terms 'discrete' and 'continuous', which we used at first (Sections III.1, 2, 3) in the (genetically) original sense, equating them with number and (linear) continuum, respectively. In the course of this chapter, however, we have almost surreptitiously changed them into 'algebraic' and 'topological', respectively. The reason for changing the terms has to be explained, and can be made clear from the history of mathematics. Thus, the former terms, which name the *'statical' entities* from classical mathematics, remind us of the times when the classical-realistic attitude still prevailed among the working mathematicians (who, by the nature of their mathematics, remained as close as possible to applied mathematics, interpreting the basic entities as qualities applying to the world). The terms 'algebraical' and 'topological' however, emphasize *operational activity* on the part of the mathematician: internal and external composition 'laws' between discrete *elements* of sets are operations of an algebraic nature, whereas topology is a 'calculus' with (open or closed) *neighborhoods of elements* of sets. The history of mathematics teaches us that, in the first half of this century, the main algebraic subjects (Algebraic Number Theory, etc. see further Section III.6) could only be further developed significantly, if suitable topologies were imposed on the structures of their interest, these topologies playing an essential but often only auxiliary role towards the solution of purely algebraic questions (e.g., the so-called Hasse-principle in the theory of quadratic forms, the local-global methods of Algebraic Geometry, etc.). Likewise the definition of suitable algebraic laws of composition of functions in topological spaces have been conducive not only to the rise of entire disciplines (Linear Analysis, Banach Algebra's, etc., see further Section III.6), but also to applications in physics (viz. through Operator Theory, Theory of Distributions, Analytic Probability Theory, etc.). To understand this historical development from the point of view of the 'duality' discrete vs. the continuous, it is convenient to note that points, discontinuities, singularities, events, chances, transformations etc. in the domain of Analysis and Topology are as much discrete entities as numbers and functions are in the domain of Number Theory and Algebra. Furthermore, on the topological side, it is imperative to note that the nineteenth century concept of continuum and continuity (e.g., of functions), in the twentieth century has been elevated to the level of topology; so as to suit the manifestly more operative and constructional nature of pure mathematics. Thus, in the topological context, a continuous function came to be an appropriately described mapping from one topological space into another. By thus operating with the concept of an 'open neighborhood' as the building block of a topological space, the classical synthetic (axiomatic) Euclidean and non-Euclidean Geometries could be tuned to topology, with the result that, nowadays, there is a tendency in mathematics education to disfavour these disciplines, even though in many applied technologies the Euclidean geometry is still a main tool of operation (comp. Bernays' remark quoted in Section III.2). Nobody, at any rate, will be ready to deny that the development of Set Theory and Systematic Topology has been mainly responsible for the stormy development of those disciplines that, in one form or another, employ the concept of a 'variety' or 'scheme'. Much less can it even be denied that the characterisation by the word 'topological' of that which in the line of classical geometry and analysis leads a derived existence, is very appropriate.

The final remark concerns the possible task of a *philosophy of mathematics*. If there is such a task, then from the complementarist viewpoint it is a modest and a limited one. Indeed, it follows from what has been said above that the implicit 'spontaneous' way (cf. Section III.1) in which mathematicians deal with the seven mentioned aspects of mathematical activity, reveals the connections and bridges that can be laid between the discrete and the continuous. Putting it in the language of Section I.1: the philosophy of mathematics necessarily employs a language which is a mixture of the natural language, a formal language or a logical language, and by that nature, is open to inexactness and contradictions (cf. Section I.7). It nevertheless may serve to glue the partial experiences of the mathematician, physicist and the general scientist together into a tentative unity. It is, most obviously, a profound study of the *history of the mathematical disciplines* that will be of great importance in this connection. As far as is known to the author, hitherto no attempts have been made to write a history of mathematics which specifically focuses attention on the relation between the discrete and the continuous. It should however be remarked that S. Bochner on several occasions has drawn attention to a duality 'discrete vs. continuous' in mathematical as well as in non-mathematical situations. Some of his examples seem to fit the scheme we have been trying to sketch in this chapter, and they will be gratefully used in the next section, even though Bochner did not seem to have intended to let these examples be supportive of the more radical view taken here.

III.6. BRIDGING THE ABYSS BETWEEN THE DISCRETE AND THE CONTINUOUS

In Table I we list under (a) those standard mathematical disciplines that are, qua subject and methods, akin to the classical theory of fields and polynomials (Galois Theory), classical algebra, representation theory and elementary theory of probability. The subjects listed under (b) are the offspring of the classical theory of (complex) functions, the classical geometries and physics. We list Logic and Set Theory on top of the division line between (a) and (b) in order to indicate their all-pervasiveness because of their normative character: they pronounce what reasonings are allowed and what is still to be called a set (cf. Section III.3). The subject Category Theory, which is about twenty years old, is listed at the bottom of the division line, because it neither is a normative subject, nor does it, as a mathematical subject, fit into any of the two columns (a) or (b). A *category* consists of all objects of a certain sort (groups, topological spaces, etc.), and of the mappings (homomorphisms, continuous maps, etc., resp.) that exist between any pair of these objects. One speaks of the category of groups, the category of topological spaces, etc., on the proviso that one identifies objects (groups, topological spaces, etc.) if they are isomorphic (isomorphic, homeomorphic, etc., resp.). A main object of Category Theory is the study of particular constructions (direct products, sums, tensor products, etc.) that can be made with objects of individual categories, in such fashion, that these constructions are being described in a universal language covering at once all categories in which such constructions are possible. In this way one may define categorically homology and cohomology theory, projective and inductive limits, etc., and these descriptions might cover categories with objects in the domain of the discrete as well as in the domain of the continuous. A powerful language has resulted that is useful and economical, because its usage makes procedures from widely diverging domains of mathematics recognisable as being analogous, thus creating problems in certain disciplines, by analogy with other disciplines. In the light of complementarism Category Theory should be rather viewed as an interdisciplinary important language device creating mathematical thinking economy, than as a discipline comparable to, say, Group Theory as it rose in the middle of the nineteenth century. The list of mathematical subjects does not pretend to be complete, but no subjects are listed the constructions and methods of which cannot be underscored by the set theory defined by A_1-A_9 of Section III.5. Further, we have as much as possible listed in one line subjects that are each other's counterpart in the two domains (a) and (b).

We repeat what it means to say that a certain subject is in one of the categories. If it is in category (a) then its underlying structure (field, space of solutions of equations, algebra, sheaf, set, etc.) is subject to algebraic operations, i.e., operations that enable us to calculate, with every two elements of the set, a third one. Epistemologically-genetically, large parts of these subjects could not be developed to their present high level purely on the basis of the discrete topology

TABLE I: *Standard mathematical disciplines*

	(a) Discrete (Algebra)	(b) Continuous (Topology)
	Set Theory Logic	
1.	Groups, Rings, Graphs, etc.	Topological Groups, Rings, etc.
2.	Algebraic Geometry	Analytic Geometry
3.	Algebraic Topology (Homological Algebra)	Analytic Topology, Manifolds
4.	Algebraic Lie Theory	Lie Theory
5.	Algebraic Sheaf Theory	Analytic Sheaf Theory
6.	Algebraic Number Theory	Analytic Number Theory
7.	Algebraic Manifolds	Analytic (& Differentiable) Manifolds
8.	Combinatorial Theory of Probability	Analytic Theory of Probability (Measure Theory)
9.	Summability Theories (Series)	Integral Calculus
10.	Difference Equations	Differential Equations
11.	Linear Algebra	Linear Analysis (Banach Algebra's)
12.	Finite Geometries	Classical Geometries
(13.	Computer Science)	
	Category Theory	

(which declares each point of their basic structures as an open and closed basic neighborhood); even though the knowledge one is after is often of a discrete nature, without reference to any sort of topology. Thus, as maybe argued from the history of these subjects, the nature of each of them and of the entities with which they deal is conducive to the definition of a topology (on their structures) that is coarser than the discrete topology (the latter being the finest topology any one set can be endowed with). These topologies usually employ for their definition those properties of the entities that make up the structures, that are quintessential to them. Whence the zero-dimensional (i.e., compact totally discontinuous) topology on (infinite) Galois groups, stemming from the fact that any infinite algebraic field extension is a direct limit of finite degree ones; the p-adic topology (p a prime number) on the field of rational numbers, using the unique division properties of the integers; the Zariski topology on algebraic varieties and algebraic groups, employing the basic properties of sets of solutions of polynomial equations; the absence of coarser topologies in Summability theories, finite group theory and finite geometry and the presence of real (complex, p-adic, respectively) topology in Euclidean algebraic topology and real (complex, p-adic, respectively) algebraic Lie theory and linear algebras.

Having thus gained a glimpse of how algebraists fetch in the hawsers thrown towards them from Topology, the question arises as to what binds the subjects under (b) together, and how they connect up with the domain of the discrete. As to the first question, it is a matter of checking through the list (b), to find that, without exception, a 'calculus of neighborhoods', i.e., a topology, inheres in any of the subjects, even though the topology in question is, historically, not made explicit in terms of the Kuratowski axioms from the outset (e.g., classical geometries, including Riemannian geometries and classical analysis originally made use of the continuum and continuous functions and the like, while only later their language became elevated to the level of topology). Also, without exception, the original versions of these subjects employed, in one form or another, what are now called topologically *complete* (real or complex) *spaces*. Except perhaps in classical geometries, most of the questions these theories are dealing with concern *functions* defined on these spaces, and it goes without saying that the classical interest is mainly concerned with *continuous* functions of this kind. It is remarkable though, that with the rise and development of the disciplines under (b), continuous functions have not lost anything of their importance, even though, nowadays, *discontinuous functions and operations* have come much more to the foreground (cf. S. Bochner). Theorems in subjects under (b), dealing with discontinuous functions and with points of discontinuity of functions, however, often seem to obtain their interest against the background of their continuous counterparts, and, at any rate, assume their meaning only in a topological context. This is one way in which in the domain of the continuous discrete entities ask for special attention. (We forego the discussion of the discrete natural constants, such as π and e, that spring from geometry and analysis and which play a wholly non-trivial role throughout all subjects under (b)). In a more extended sense though, singularities, chances, events, 'particles', transformations, points, etc. (cf. Section III.5), in the context of the subjects (b), are unthinkable outside of the topological context in which they are defined.

But now for the *algebraic operations* in the domains under (b). Whereas, generally, the subjects under (a) are of an algebraical nature, topological operations added as means to an algebraic end, the subjects under (b) can be generally said to be of a topological nature, the algebraic operations being added to serve the function-theoretical and topological ends. In both cases, the algebraic and topological operations usually are preferred to be defined in such a manner that they concur with each other in this sense, that one requires the algebraic operations – considered as 'functions' on the two-fold product of the underlying structure with values in the same structure – to be *continuous*. This circumstance (viz., that in both domains algebraic operations are usually continuous), together with the axiomatic-formalist set-up of the disciplines, often shrouds the different epistemological origins of the sub-disciplines (in a manner analogous to people facing a bridge across a deep ravine: from out of which side has the bridge been built?). The following example may clarify this further. It concerns all subjects, large chunks of which consist of the application of standard function theory or Analysis, on complete topological spaces (e.g., algebraic and analytic Lie theory, Banach Algebra, etc.). 'Abstract' Analysis on these spaces then, may be set up in such axiomatic fashion that no special assumptions on the nature of the underlying space are being made. One may thus derive a considerable body of general theory without paying attention to the fact that in a later stage the theory splits into two major, totally different, parts viz., the non-Archimedean function theory (on a totally disconnected, or totally discontinuous base space) and the Archimedean theory (on a base space with Archimedean norm). Whereas in most applications the latter theory concerns spaces that have underlying topologies derived from the natural topology of the *real numbers*, the former theory employs topologies that in one form or another are derived from *algebraic properties* of certain of its subsets (such as for example, divisibility properties of integers or of polynomials in the theory of local fields). Indeed, non-Archimedean norms give rise to a theory of convexity and to a Linear Analysis that looks wholly different from real or complex Linear Analysis, and their major applications take place in the domain of Number Theory, the theory of Algebraic Groups (hence in Algebraic Geometry) and algebraic Lie theory.

The question which now arises is: if the totally discontinuous topologies are so preponderant in algebraic subjects, then why does one not replace the headings of the columns (a) and (b) by 'totally discontinuous' and 'continuous', respectively? There are many arguments against such a replacement which we will not all consider (e.g., the Zariski topology in Algebraic Geometry is not totally discontinuous, etc.). The main objection we have to such a proceeding is, that then the genetical origin of the subjects would again be shrouded, because it is the basic wish of complementarism to keep track of the historical-architectural make-up of the edifice of mathematical knowledge: the word 'discrete' is not meant to be a topological word, and even if it is taken as such, one may realise that the so-called 'discrete topology' on a set is not really a topology.

A second point we have promised to deal with is the examples adduced by Bochner to support his view that something of a 'duality' discrete vs. continuous shines through in mathematics. He even conjectures that, some day, it may be that such a kind of duality appears as a central pronouncement in mathematics, analogously to the principle of complementarity by Bohr. Naturally, we do not say that the complementarity 'principle' we have been defending should be viewed as *the* central pronouncement Bochner is looking for, but we do certainly submit it as a candidate. If we do understand Bochner well enough then the duality he has in mind should be a very strict one, reflecting itself in formulas rather than in disciplines, as is the case in complementarism. For instance, he mentions that in the Poisson summation formula the left hand side (a sum) is a discrete device and the right hand side (an integral) is a continuous device. Similar things can be said of the relation between a periodic function (continuous) and its set of (discrete) Fourier coefficients; of the relation between the set of eigenfunctions of a Sturm-Liouville equation and the set of eigenvalues; of Cauchy's residue formula (calculating the discrete sum of the residues of a function in terms of an integral over a closed curve) and of the duality between homology (continuous) and cohomology (the discrete). If we look well at these examples then we see that, except for the last one, they all concern the subject No.9 of our list: Summability theory vs. Integral Calculus. Indeed, in our previous discussions we have been hardly paying attention to that subject, preoccupied as we were with those subjects that have underlying spaces that lead a derived existence. The evaluation of sums (of numbers or functions, for example) usually takes place at a much more original level and it is the central problem of Integral theory to convert sums into integrals. This it achieves usually by letting integrals be limits of sums, and the remarkable feat of the 'dualities' mentioned above is that sums of values of certain functions can be equated with integrals of suitably defined functions that can be derived from the summands of the sums in question. This leaves us with the final example on the (co)homology. The (co)homological operations the attention is drawn to have a typical place in subjects where varieties, schemes and manifolds stand in the center of interest. In a very wide sense cohomology operations break up a (topological) manifold (variety, etc.) by inscribing symplices in them, in a manner, roughly speaking, in which one can inscribe a polygon inside a circle. This is a discrete operation applied to something continuous (the topological space in question). Operations with the set of all symplices that can be inscribed in the manifold lead to vital information on the structure of the manifold. Quite dually but roughly, homological operations measure the nature of the continuity structure of the manifold, by performing continuous operations (the so-called contraction) on closed curves going through points of the manifold. For great classes of manifolds the homological structure is known once the cohomological structure is, and vice versa. We do not see, however, how this example fits into the scheme of subjects (a) and (b), simply because there are homology and cohomology theories for subjects under (a) *and* under (b), these theories being analogous for subjects in the same line, but different, as the definition of the (co)-homology in corresponding subjects is not usually the same because the topologies for them differ considerably. Hence, only in the examples from Integral theory the said dualities are examples of bridges spanning the abyss between discrete operations and continuous operations. So it may be that, after all, mathematicians may agree that the 'duality' we have been trying to point out is a fundamental one concerning *mathematical disciplines*, revealing something non-trivial about their epistemological-genetical origin.

Selected Bibliography

Angelelli, I., *Studies on Gottlob Frege and Traditional Philosophy*, D. Reidel, Dordrecht, 1970.

Archimedes, *The Works of*, ed. by T. L. Heath, New York, Dover, 1897.

Aristotle, *The Basic Works of*, ed. by R. McKeon, New York, Random House, 1941.

Benacerraf, P. and Putnam, H. (eds.), *Philosophy of Mathematics. Selected Readings*, Prentice-Hall, Englewood Cliffs, 1964.

Beth, E. W., *Inleiding tot de Wijsbegeerte der Wiskunde*, Dekker en van de Vegt, Nijmegen, 1940.

Beth, E. W., *The Foundations of Mathematics*, North-Holland, Amsterdam, 1965.

Beth, E. W., *Moderne Logica*, Assen, Van Gorcum, 1969.

Beth, E. W. et Piaget, J., *Epistemologie Mathématique et Psychologie*, Presses Universitaires de France, Paris, 1961.

Bishop, E., *Foundations of Constructive Analysis*, McGraw-Hill, 1967.

Bochner, S., 'Analysis on Singularities', in: *Complex Analysis*, 1972, Vol. II, Proc. Conf. Rice Univ., Houston, Texas, 1972, Rice Univ. Studies 59, 1973, No. 2, pp. 21-40.

Bourbaki, N., 'L'architecture des Mathématiques', in *Les Grands Courants de la pensée mathématique*, Cahier du Sud, Paris, 1948.

Brouwer, L. E. J., *Over de Grondslagen der Wiskunde*, Maas en Van Suchtelen, Amsterdam, 1907.

Brouwer, L. E. J., 'Intuitionism and Formalism', in *Philosophy of Mathematics*, ed. by P. Benacerraf and H. Putnam.

Bunge, M., *Intuition and Science*, Prentice-Hall, Englewood Cliffs, 1962.

Cantor, G., *Gesammelte Abhandlungen*, Georg Olms Verlagsbuchhandlung, Hildesheim, 1966.

Carnap, R., 'The Logicist Foundations of Mathematics', in *Philosophy of Mathematics*, ed. by P. Benacerraf and H. Putnam.

Chomsky, N., *Syntactic Structures*, Mouton, 's-Gravenhage, 1957.

Chomsky, N., *Language and Mind*, Harcourt, Brace and World, New York, 1968.

Cohen, P. J., *Set Theory and the Continuum Hypothesis*, W. A. Benjamin, Inc., New York, 1966.

Davis, M., *Computability and Unsolvability*, McGraw-Hill, New York, 1958.

Deligne, P., 'La Conjecture de Weil I', *Publ. Math.* **43**, Inst. des Hautes Etudes Sc., Paris, 1974, pp. 263-307.

Descartes, R., Selections from *Rules for the Direction of the Understanding* (written between 1619 and 1628), in *Philosophers Speak for Themselves. From Descartes to Locke*, ed. by T. V. Smith and M. Grene, Univ. of Chicago Press, Chicago, 1940.

Descartes, R., *Geometry* (1637), Dover, New York, 1954.

Descartes, R., *Discourse on Method* (1637), Liberal Arts Press, Indianapolis, 1960.

Descartes, R., *Meditations* (1641), Liberal Arts Press, Indianapolis, 1960.

Dieudonné, J., 'The Weil Conjectures', *The Mathematical Intelligencer* **10**, 1975, The Yellow Press, Springer-Verlag, Heidelberg, quarterly.

Freudenthal, H., 'Qualität und Quantität in der Mathematik', *Euclides* 8, 1932, pp. 89-98.

Heath, T. L., *A History of Greek Mathematics*, Clarendon Press, Oxford, 1921, 2 volumes.

Heyting, A., *Intuitionism. An Introduction*, North-Holland, Amsterdam, 1956.

Hilbert, D., 'On the Infinite', in *Philosophy of Mathematics*, ed. by P. Benacerraf and H. Putnam.

Hofman, J. E., *Leibniz in Paris* (1672-1676), His Growth to Mathematical Maturity, Cambridge Univ. Press, 1974.

Hurd, A. and Loeb, P. (eds.), *Victoria Symposium in Nonstandard Analysis*, L.N.M. No. 369, Springer-Verlag, 1974, Berlin.

Kamke, E., *Mengenlehre*, Walter de Gruyter, Berlin, 1965.

Kant, I., *Critique of Pure Reason* (1781, 1787), St. Martin's Press, New York, 1929.

Kant, I., *Prolegomena to Any Future Metaphysics* (1783), Liberal Arts Press, Indianapolis, 1950.

Kleene, S. C., *Introduction to Metamathematics*, North-Holland, Amsterdam, 1952.

Körner, S., *The Philosophy of Mathematics. An Introduction*, Harper, New York, 1960.

Lakatos, I. (ed.), *Problems in the Philosophy of Mathematics. Proceedings..*, North-Holland, Amsterdam, 1967.

Lakatos, I. (ed.), *The Problems of Inductive Logic*, North-Holland, Amsterdam, 1968.

Lakatos, I. and Musgrave, A. (eds.), *Problems in the Philosophy of Science*, North-Holland, Amsterdam, 1968.

Leibniz, G. W., 'Necessary and Sufficient Truths', in *Philosophers Speak for Themselves*, ed. by T. V. Smith and M. Grene, Univ. of Chicago Press, Chicago, 1940.

Leibniz, G. W., 'Beobachtungen über die Erkenntnis, die Wahrheit und die Ideen' (1684), in *G. W. Leibniz. Hauptschriften zur Grundlegung der Philosophie*, ed.

by E. Cassirer, Felix Meiner, Hamburg, 1904, 2 volumes.
Leibniz, G. W., 'Specimen Dynamicum' (1695), in *Hauptschriften*, ed. by E. Cassirer.
Leibniz, G. W., 'Zur Prästabilierten Harmonie' (1696), in *Hauptschriften*, ed. by E. Cassirer.
Leibniz, G. W., 'Betrachtungen über die Lebensprinzipen und über die Plastischen Naturen' (1705), in *Hauptschriften*, ed. by E. Cassirer.
Leibniz, G. W., 'Die Monadologie' (1714), in *Hauptschriften*, ed. by E. Cassirer.
Leibniz, G. W., 'Aus den "Metaphysischen Anfangsgründen der Mathematik" ' (1714), in *Hauptschriften*, ed. by E. Cassirer.
Luxemburg, W. A. J., 'What is Non-Standard Analysis?', *American Math. Monthly* **80**, June-July 1973, part II, pp. 38-67.
Mendelson, E., *Introduction to Mathematical Logic*, Van Nostrand, Princeton, 1965.
Mooij, J. J., *La Philosophie Mathématique de H. Poincaré*, Cahiers mathématiques, Paris, 1969.
Morris, Ch. W., *Foundations of the Theory of Signs*, Chicago, 1938.
Nagel, E. and Newman, J.R., *Gödel's Proof*, N.Y. Univ. Press, 1958.
Newton's *Philosophy of Nature. Selections from his Writings*, ed. by H. S. Thayer, Hafner, New York, 1953.
Nivette, J., *Grondbegrippen van de generatieve Grammatica*, Brussel, Ninove, Steppe, etc. 1970.
Piaget, J., 'Logique et connaissance scientifique', in *Encyclopédie de la Pleiade*, ed. by J. Piaget, 1969.
Plato, *The Collected Dialogues*, ed. by E. Hamilton and H. Cairns, Bollingen Foundations, New York, 1961.
Poincaré, H., *Science and Hypothesis* (1902), Dover, New York, 1952.
Poincaré, H., *The Value of Science* (1905), Dover, New York, 1958.
Poincaré, H., *Science and Method* (1908), Dover, New York, not dated.
Poincaré, H., *Mathematics and Science: Last Essays* (1913), Dover, New York, 1963.
Quine, W. V. O., *From a Logical Point of View*, Harper, New York, 1963.
Robinson, A., *Non-Standard Analysis, Studies in Logic and the Foundations of Mathematics*, North-Holland, Amsterdam, 1966.
Rosser, J. B., *Logic for Mathematicians*, McGraw-Hill, New York, 1953.
Rosser, J. B., *Simplified Independence Proofs*, Acad. Press, New York, 1969.
Russell, B., *An Essay on the Foundations of Geometry* (1897), Dover, New York, 1956.
Russell, B., 'Logical Atomism', in *Logical Positivism*, ed. by A. J. Ayer, The Free Press, New York, 1959.
Russell, B. and Whitehead, A. N., *Principia Mathematica*, Vol. 1 (1910), Cambridge Univ. Press, Cambridge, 1927.

Serre, J. P., 'Valeurs propres des endomorphismes de Frobenius', *Sém. Bourbaki* **26**, No. 446, 1973/74; L.N.M. No. 431, Springer-Verlag, Berlin, pp. 190-204.

Suppes, P., *Introduction to Logic*, Van Nostrand, Princeton, 1964.

Tarski, A., *Logic, Semantics, Metamathematics*, Clarendon Press, Oxford, 1956.

Tarski, A., 'The Semantical Concept of Truth and the Foundations of Mathematics', in *Phil. and Phen. Research* **4**, 1944; Dutch Translation by E. W. Beth in *Euclides* **30**, 1954/55.

Tarski, A., 'Truth and Proof', in *Scientific American*, June 1969, pp. 63-77.

Tarski, A., *Introduction to Logic*, Oxford Univ. Press, New York, 1965.

Weil, A., *Bulletin of the American Math. Society* **81**, No. 4, pp. 676-688, review of J. E. Hofman's work on Leibniz in Paris.

Weyl, H., *Gesammelte Abhandlungen*, Springer, Berlin, 1969.

Wheeler, N., *Einstein's Vision*, Springer, Berlin, 1970.

Wilder, R. L., *Introduction to the Foundations of Mathematics*, John Wiley & Sons, New York, 1965.

References for Further Study

Whereas the text of this book refers directly to the Selected Biography, the bibliography below may be helpful for the student who wishes to study the different aspects of the Foundations of Mathematics, touched on in the text, further. It is arranged chapterwise.

CHAPTER I (Logic, etc.)

Bartley, W. W., 'Lewis Carroll's Last Book on Logic', *Scientific American* **227**, July 1972.

Bell, J. L., Slomson, A. P., *Models and Ultraproducts*, North-Holland, 1969.

Bochenski, I. M., *Ancient Formal Logic*, North-Holland, 1968.

Curry, H. B., *Outlines of a Formalist Philosophy of Mathematics*, North-Holland, 1951.

Van Dalen, D., Doets, E. and De Swart, J., *Sets*, Pergamon Press, 1977.

Freudenthal, H. *et al.*, *The Concept and the Role of the Model in Mathematics and Natural and Social Sciences*, Reidel, Dordrecht, 1961.

Freudenthal, H., *The Language of Logic*, Elsevier, Amsterdam, London, New York, 1966.

Gödel, K., *The Consistency of the Axiom of Choice and of the Generalized Continuum Hypothesis with the Axioms of Set Theory*, Princeton, 1940.

Gödel, K., *The Consistency of the Continuum Hypothesis*, Princeton, 1940.

Gödel, K., 'What is Cantor's Continuum Problem?' *Am. Math. Monthly* **54** (1947).

Mostowski, A., Tarski, A. and Robinson, R. M., *Undecidable Theories*, North-Holland, 1953.

Nagel, E., 'Logic without Ontology' in *Naturalism and the Human Spirit*, ed. by Y. H. Krikorian, Columbia Univ. Press, 1944.

Quine, W. V. O., *Methods of Logic*, revised edition, New York, 1959.

Robinson, A., 'Standard and Nonstandard Number Systems, The Brouwer Memorial Lecture', *Nieuw Archief voor Wiskunde* **21**, No. 2, 1973.

Rubin, H. and Rubin, J., *Equivalences of the Axiom of Choice, Studies in Logic*, North-Holland, 1970.

Scholz, H. and Hasenjaeger, G., *Grundzüge der mathematischen Logik*, Berlin-Göttingen-Heidelberg, 1961.

CHAPTER II (History, etc.)

Apostle, H. G., *Aristotle's Philosophy of Mathematics*, Chicago, 1952.
Bochner, S., *The Role of Mathematics in the Rise of Science*, Princeton, 1966.
Bourbaki, N., *Eléments de l'histoire des mathématiques*, 2me édition, Hermann, Paris, 1969.
Van Dalen, D. and Monna, A. F., *Sets and Integration. An Outline of the Development*, Groningen, 1972.
Frege, G., *Die Grundgesetze der Arithmetik*, Hildesheim: Olms, 1962.
Freudenthal, H. and Heyting, A. (eds.), *The Collected Works of L. E. J. Brouwer*, 2 volumes, North-Holland 1976/1977.
Gödel, K., 'What is Cantor's Continuum Problem?' *Am. Math. Monthly* **54** (1947).
Heath, T. L., *A Manual of Greek Mathematics*, Dover, 1963.
Kleene, S. C. and Vesley, R. E., *Introduction to Intuitionistic Mathematics*, North-Holland, 1965.
Kline, M., *Mathematical Thought from Ancient to Modern Times*, Oxford Univ. Press, 1972.
Kreisel, G., 'Hilbert's Programme', *Dialectica* **12** (1958).
Lukasiewicz, J., *Aristotle's Syllogistic, from the Standpoint of Modern Formal Logic*, Oxford, 1951.
Neugebauer, O., *Vorgriechische Mathematik*, 2nd edition, Springer-Verlag, 1969.
Von Neumann, J., 'Die formalistische Grundlegung der Mathematik', *Erkenntnis* **2** (1931).
Rényi, A., *Dialogues on Mathematics*, Holden-Day, 1965.
Struik, D. J., *A Concise History of Mathematics*, Dover, 1948.
Van der Waerden, B. L., *Science Awakening*, Noordhoff, Groningen, 1973.
Weyl, H., *Philosophy of Mathematics and Natural Science*, Princeton, 1949.

CHAPTER III (Philosophy and Mathematics, etc.)

Bar-Hillel, Y. and Fraenkel, A. A., *Foundations of Set Theory*, North-Holland, 1958.
Bernays, P. and Fraenkel, A. A., *Axiomatic Set Theory*, North-Holland, 1958.
Bochner, S., *Eclosion and Synthesis*, Benjamin, New York, 1970.
Cohen, P. J. and Hersch, R., 'Non-Cantorian Set Theory', *Scientific American* **217**, December 1947.
Van Dalen, D. and Monna, A. F., *Sets and Integration, An Outline of the Development*, Groningen, 1972.
Freudenthal, H., *Mathematics as an Educational Task*, D. Reidel, Dordrecht, 1973.
Freudenthal, H. *et al.*, *The Concept and the Role of the Model in Mathematics and Natural and Social Sciences*, Reidel, Dordrecht, 1961.
Halmos, P., *Naive Set Theory*, New York, 1960.

Mac Lane, S., *Categories for the Working Mathematician*, G.T.A., Springer-Verlag, Berlin, 1971.

Von Neumann, J., 'The Mathematician', in *Collected Works*, Pergamon Press 1961, pp. 1-9, or (the German translation) in Michael Otte etc.

Otte, M. (ed.), *Mathematiker über die Mathematik*, Springer-Verlag, 1974.

Piaget, J. *et al.*, *Logique et connaissance scientifique*, Encyclopédie de la Pléiade, Editions Gallimard, Paris, 1967.

Steen, L. A., 'New Models of the Real Number Line', *Scientific American* **225**, August 1971.

Weyl, H., *Das Kontinuum*, Chelsea, New York, 1932.

Wheeler, J. A., *Geometrodynamics*, Acad. Press, New York, 1962.

Index of Names

The chapter number indicator is not repeated.

Index of Subjects

The chapter number indicator is not repeated